Life Is Funny

A Riveting Tale of Comedy, Hairdressing and Texas Politics

Leigh Anne Jasheway-Bryant

AF699460

Comedy Workout Publishing
Eugene, OR

A Comedy Workout Trade Paperback Original

Copyright © 2007 Leigh Anne Jasheway-Bryant

All rights reserved. No part of this book may be used or reproduced in any form without written permission except in the case of reprints in the context of reviews. For information, call The Comedy Workout at 541/688-1674 or e-mail at lajfun@accidentalcomic.com.

Published in the United States

ISBN 0-9674486-7-0

To

Jody, you know who you are,
but who knows where you are.
And if you're thinking of running for office,
please read this book first.

Nancy for creative editing and being able to take
(and improve) a joke.

Brad, Maddy Lou, Justin Time, and Penny for being funny
and silly and unlike the voices in my head,
at least partially real.

Anyone who has ever encouraged me, even by laughing at
me while waiting in line at the ATM. This is all your fault.

Also by Leigh Anne Jasheway-Bryant:

Are You Playing with Me?
Bedtime Stories for Dogs
Bedtime Stories for Cats
The Rules for Dogs
The Rules for Cats
Don't Get Mad, Get Funny
If I Was a Dog, I'd Be a Better Person
I'm Not Getting Older (I'm Getting Better at Denial)
Laugh Lines are Beautiful
Serious Side Effects
Yoga for Your Funny Bone

No Time for Trouble

You could say this about Judy – the girl could talk. She had once yammered on to a customer for a solid twenty minutes without realizing the woman was sound asleep. Judy did pause for a few seconds when the old broad's head rolled forward onto her chest and it seemed she might be dead, but Nan's electric clippers jolted them both back to life like a remote defibrillator. The woman's face was a little blue when she woke up, but that was just the reflection of her hair. And Judy picked up her story without missing a beat.

Today however, as the neon sign brashly proclaimed *Lone Star Cuts – Home of Texas' Biggest Hair*, Judy wasn't her usual chatty self. She grabbed a comb, ran her thumb over the edge, and looked around the shop. It was always a shame to waste her stories on the likes of Rae Ann, who wasn't a big laugher before the facelift. Now that she was apparently under doctor's orders never to display any kind of movement above the neck, it was a damned shame. Rae Ann had recently learned how to smile without smiling, frown without frowning, and look surprised simply by blinking rapidly. Judy might as well have tried to entertain the residents of Madame Tussaud's Wax Museum.

At least Judy wasn't alone in her bad mood. Everyone in the salon was about as perky as the half-dead Christmas tree in the corner next to the window. April, Lone Star's owner, kept promising to take the poor thing to the curb so it could be picked up by the Recycling Guys – sweaty young men the staff and clients all loved watching through the front window. But it was Thursday and the Guys only showed up on Tuesdays between 10-ish and 1-ish, so the tree just sat there, as sad and unwanted as the hair that piled up beneath the stylists' chairs like little hair wreaths. It was all very festive in a *Week After the Grinch Stole Christmas* way.

April bore full responsibility for the comatose tree. It was she who insisted on putting up a fresh tree the day after Thanksgiving every year and then procrastinating about its removal until it was just a needle away from spontaneous combustion. April thought the longer the tree stayed up and decorated, the longer the holiday cheer would last. Although a dead tree could only be cheerful to people who had upped their meds in mid-December.

There was a definite feeling of *"I'm glad the holidays are over, but now what's there to look forward to besides bills I can't pay and New Year's Resolutions I'm already planning on breaking"* in the air today. The only one who was excited to be alive was Leon – Lone Star's only male stylist – for he was born with the football gene and couldn't help flashing a shit-eating grin whenever the Longhorns or the Aggies or the Cowboys were having a good year. He didn't care who won as long as they were Texas boys. Unless they were Texas boys playing for Oklahoma. That just wasn't right. Not to mention that football meant cheerleaders and cheerleaders made life worth living in Leon's not-so-humble opinion.

Judy especially hated the long slow days after the holidays, preferring the salon when it was buzzing with women getting ready for their holiday parties. Since Lone Star Cuts was only six blocks from the Capitol Building, most of the women who came in were well-heeled and well-connected. To the bank. As much as politics got Judy's dander up, she liked the women's stories and she loved their tips. If every week were like the week before Christmas, Judy could retire in five years. Unfortunately, every week was more like today.

As Judy started to comb out Rae Ann's wet hair, she shot a glance at Donna sitting at the front desk. She could count on Donna to be quick with a nice round, boisterous laugh that should have come from a round, boisterous woman, not one so tiny she had to buy her clothes in the Preteen Department at JC Penney's. Watching that laugh come from Donna's tiny pursed mouth was like watching a Basset Hound bark coming from a Teacup Poodle, which on this day was exactly what Donna's hair reminded Judy of.

Even with Donna within earshot, Judy preferred a crowd, so for once, she worked in silence. Unfortunately, silence wasn't Judy's friend. Too much quiet in her overactive brain gave her time to think and thinking often lead to trouble. Today's troublesome thoughts were all about Juan. He had been popping into her mind a lot lately. It was the holidays – there was nothing like the whole *peace on earth goodwill to men* nonsense to conjure up the most depressing moments of one's life.

Transfixed, Judy stopped combing and watched as the three years of their relationship flashed by like outtakes from a bad movie. Juan yelling at her in Spanish even though he knew she only understood about one out of every five words. The whole relationship would have been better with subtitles. Cut. Judy telling Juan to get out of her life after he threatened to kill one of the cats because it had peed on the sofa. Cut. Juan jimmying open her trailer window and standing there watching her as she slept until she woke up, screamed, and threatened to call the cops. Whenever she replayed this scene, Judy could hear the theme song from *Jaws*. Cut. The voice on the other end of the phone, almost a year ago, telling her that

Juan had been killed in a car wreck on the highway to Tomball. Not surprisingly, there had been two empty six-packs on the floor. Cut. Sad ending and no popcorn. Cut.

Judy shook her head, trying to get the memories to fly back to wherever they'd come from. *When I'm famous,* she thought, I'll rewrite the ending. *Oh, hell, I'll rewrite the whole story. Make it funny. Juan ends up being swept up by a twister on his way to the Stop 'N Rob and finds himself plopped down on the roof of a halfway house. Antonio Banderas stars in the role...*

"Judy. Judy!"

It was Rae Ann's voice. The brain fog started to clear. Antonio's smiling face and suggestively-cocked eyebrow morphed into Rae Ann's impassive face and motionless brows.

"Are you gonna cut my hair or stare off into space all day?" Rae Ann demanded.

Called back to what passed for real life, Judy slowly began cutting Rae Ann's hair. Everyone knew Judy wasn't the fastest or the best hairdresser in town, but she was the funniest. Being funny was both a blessing and a curse: a blessing because it brought in customers, which meant that Judy didn't have to panhandle on Guadalupe Street near campus in order to survive. A curse because Judy had discovered early on that big girls who don't just sit back and let you make jokes about them, but who actually make better jokes back, were too much for most guys to handle. Except, apparently, choice specimens like cocaine addicts and cat-hating alcoholic stalkers who don't speak English.

Judy cut a little too much off the right side of Rae Ann's hair. "Oops!"

Rae Ann quickly turned her head towards the mirror to see if there was any noticeable damage. Not this time. "Dahling, one more 'Oops!' and I'm gonna have to find another beautician or at least a good cardiologist," she said, catching Judy's over-mascaraed eyes with her own and holding them like an alpha bitch establishing her dominance. Judy didn't even squirm.

"That's what you always say, Rae Ann, but I know better." Judy broke eye contact first, calling it a draw, and went on snipping, taking another half inch off the left side to even things out. "You're such a bad liar," Judy said, "I bet you can't even fake an orgasm properly!"

Rae Ann rose a full two inches in her chair and looked as indignant as her recent plastic surgery would allow. It can be hard getting a good scowl going when your eyebrows won't move on their own.

From the desk, Donna caught her breath. *Not again*, she thought. Judy was always pushing clients to the edge. One of these days she was going to scare off someone really important. Someone like Rae Ann, who owned three car dealerships on the edge of town and could easily afford to go somewhere else to get her big hair fluffed and lifted twice a month. Somewhere they didn't employ stylists who brought up the word "orgasm" just for shock value.

"Honey, I could teach *Fakin' Orgasm 101* at Austin Community College!" Rae Ann sputtered indignantly, making sure her lips didn't actually move, like a ventriloquist with a dummy gone AWOL. Donna breathed a sigh of relief and stifled a giggle. Stifling wasn't in Judy's DNA, so she slapped her thigh and laughed until tears rolled down past her tarantula-like eyelashes and onto her cheeks. Thank God for waterproof mascara. If she could afford it, Judy would have bought the stuff in 55-gallon drums.

"That's one class I do not need to take!" Judy had not only faked orgasms, she'd faked interest. Lately, she'd been faking it by herself. Now that was sad.

"Don't go assumin' just cuz ah'm richer than God ah don't know a few things about the world," Rae Ann said, drawling for effect. "If memory serves, mah best friend taught me how to fake it the night ah had my comin' out party."

"I bet that was some party!" Judy laughed again. "Aren't you just the little devil?"

"I've had my moments," Rae Ann declared proudly.

Judy nodded at her client. "I'm sure you still do." She took a final snip of hair, then started combing it out. You never could tell what was really going on under the big hair of a rich Texas woman. She handed Rae Ann a mirror.

Rae Ann grabbed the handle and circumnavigated her hair, looking for "don'ts" in her "do." She didn't find any. It looked fine. For a change.

"You done good, hon. 'Bout time. Ah was fixin' to start cuttin' it myself!"

Judy faked a gasp. "You wouldn't. You know what they say, 'The woman who cuts her own hair has friends who laugh behind her back.'" She crossed her purple-tipped fingers over her chest and looked at her client approvingly. Judy hoped she'd look as good as Rae Ann when she was in her 60s. Or was it 70s? It was hard to tell. Rae Ann had had so much plastic surgery, there was probably a recycling code behind her ear. Although Judy hadn't found it yet.

Rae Ann ran her fingers through her tresses. Suddenly overcome by an urge, Judy grabbed the scissors once again and made a final snip. Rae Ann just shrugged her shoulders and sighed. Down went the scissors again. "Remember," Judy interjected, "Any time you want to adopt me, just give me a call. You know where I'll be."

This was Judy's running joke with all her rich clients. She figured she had a better chance of being adopted by a millionaire client than winning the lottery (which was at $144 million today. Judy had her usual one ticket with the numbers 7, 11, 12, 27, 33, 39, 45 – Nolan's age, Stevie Ray's age, the dress size she'd like to diet down to, the date her divorce became final, her age, her favorite number, and the number of calories in a bite of her favorite pecan pie – taped to her mirror.)

She loosened the smock from Rae Ann's neck and whipped it off with a flourish, almost yanking out one of her client's diamond snowman earrings in the process. As Rae Ann rubbed her ear between her fingers, Judy fastened and unfastened the Velcro strip that held the smock together at the neck. Judy had a thing for Velcro. In high school, she'd sneak in after the skinny girls had finished gymnastics practice and un-Velcro the tumbling mats. There was something about that ripping noise that just got to her. Last year for her birthday, the girls at the salon had given her an anatomically-correct, life size replica of her ex-husband, with all his body parts velcroed on. When she was feeling depressed, she pulled them off one at a time, saving, of course, the best for last.

"Are you sure you don't want me to blow it dry?" Judy inquired tauntingly, picking up a blow dryer in each hand, like she was challenging Rae Ann to a showdown at a few minutes past high noon.

Rae Ann stood up, brushed a few stray hairs off her ample chest – which was real, judging from the way it sagged, one side slightly lower than the other, like an undescended testicle – and placed one heavily ringed hand on her bony hip. "How long have I been coming in here?"

"Four, five years?" Judy guessed. She looked over at Donna, who shrugged. That was before her time.

"Five years in March," Rae Ann interjected. "And how many times have I let you blow my hair dry?"

"One." Judy had no trouble remembering. It was Rae Ann's first visit and the blow dryer had shorted out. There had been a little fire, but Judy had been able to put it out by spritzing it with water before it jumped from the top of Rae Ann's head to the sides and back. "But didn't I get those flames out before they damaged your perfect eyebrows?" Judy whined. She holstered the blow dryers.

Rae Ann smiled, her lips turning up almost unnoticeably. "I'll let it air dry, thanks."

"Okay, but if you catch pneumonia and die, don't blame me!" Judy said in mock horror. "Just make sure you put me in the will first."

Rae Ann handed Judy a $20 bill, all the cash she had on her. She didn't like to carry cash because Lord knows where it had been. She carried plastic. Platinum plastic. "Don't spend it all at Nieman Marcus."

"I wouldn't dare. I couldn't disappoint the clerks down at Goodwill. Besides, after buying my boys video games and that damned Habitrail for Christmas, I won't be spending any money for eight, nine years." When it came to disposable income, Judy's was permanently indisposed. She took the twenty and tucked it away in her tip jar, where it stood out among three ones, two quarters and a note that said "Never accept a ride from a stranger wearing nothing but a trench coat."

Rae Ann walked over to the front desk and settled her bill with Donna. "Give yourself five dollars, dear," Rae Ann said generously. Donna unhesitatingly did. Rae Ann was one of a few customers who tipped for just adding up the bill. Donna liked that about her.

Judy began sweeping up her station. Rae Ann snapped her purse shut and sashayed towards the door, her fragile-looking ankles holding up remarkably well in her high, pointed heels. Her doctor had tried to talk her into wearing orthopedic shoes the last time he removed a bunion, but she told him she'd rather die. A woman has to do what a woman has to do.

The phone rang. Donna picked it up just as Rae Ann opened the door and the chimes began their incessant rendition of "The Eyes of Texas Are Upon You." Sometimes Donna could barely hear the person on the other end of the phone. She'd been begging April for weeks to turn down the volume. Maybe she'd just have to get a screw driver and fix that damn chimes herself.

"See ya'll in a month," Rae Ann shouted above the chimes. "And Judy, you stay out of trouble."

Judy smiled. "Me, trouble? I don't have time for trouble."

"Can't you get this doorbell to play something else?" Rae Ann yelled over her shoulder as the door closed slowly behind her. "Yellow Rose of Texas, Ah Want to Go Home With the Armadillo?"

"I've been lobbying for 'Mamas Don't Let Your Babies Grow Up to Be Bald Guys,' but April likes this little ditty," Judy yelled back. By that time, however, Rae Ann was already in the parking lot, standing at the door to her gold Beemer, drumming her impatient but perfect nails on the roof as she decided whether to go directly to her massage therapist or make her semi-weekly unannounced visit to Garrett's office to make sure he wasn't up to anything. Ever since last year's affair with that young law clerk, she'd decided he deserved a little more anxiety in his life.

Inside, Judy checked the appointment calendar over Donna's shoulder. *A break*, she thought as she plopped down in her chair, exhausted and desperately in need of another cup of coffee. Unfortunately, she was trying to cut back. Her doctor had told her it might be coffee that was making her breasts lumpier than usual.

Judy looked over at Donna, who was erasing something from the schedule and blowing eraser crumbs on the floor.

"What do you have planned for New Year's Eve?" Judy asked with a yawn.

"Oh, nothing exciting. Think I might re-paper the bathroom."

Judy picked up the hand-held mirror and stared at her face. In this light, her pores were the size of potholes. "Just the traditional celebration, huh?" She put the mirror down with disgust. "You've been saying you're gonna re-paper that bathroom for months now. Your cat could have done the job quicker."

"Well, I tried to get him to do it, but George Walker kept shredding the strips when he was applying the glue. Damn cat!"

They both laughed. "That's what happens when your cat is named after a lying scumbag. You're lucky he hasn't declared war on the downstairs neighbors."

Donna, a lifelong Democrat and proud of it, had had the misfortune of adopting a cat that had come with a name. It was on his collar at the Humane Society. She'd tried calling him dozens of other names – Fluffy, Smoky, Tinkerbell, etc. – but he'd only come when she yelled George Walker.

"You got any plans?" Donna asked, hoping maybe Judy would invite her to tag along to a movie or something, or at least give her an idea of something to do besides sit home and de-flea the cat on the biggest party night of the year.

"The ex has got the boys 'til Sunday, so Mark and I will probably go dancing." Judy hadn't told Mark about the dancing yet. "Then we'll go home and I'll see if I can get him to feed me grapes one at a time while massaging my feet."

"Mark? The one with the tattoo?"

Judy glared at her. "No, that was Mike. I kicked him and his 'Giddy Up' tattoo out ages ago. Keep up!"

"Sorry!"

"This one is Mark. I've been seeing him for three weeks... no, almost a month. No tattoo, at least that I've noticed. But I haven't looked everywhere."

Judy liked Mark. She liked them all in the beginning, although Mark did seem different. For one thing, he had a real job. He was a research analyst for a polling firm. Judy wasn't even sure what that meant. But she knew it meant he wasn't collecting unemployment and he didn't have time to spend all day at a bar. And unlike most of the guys she had dated, who didn't want anything to do with Nolan and Stevie Ray because they wanted to be boys themselves and couldn't stand the competition, Mark took to her sons right away. He'd even bought them a Christmas present, although another rat wasn't Judy's idea of the ideal holiday gift. Rats made Judy anxious, perhaps because they reminded her of her choices in men. They had beady little eyes and pooped everywhere. The rats and the men. But a rat is what her sons wanted and a rat is what Mark got them. And now she'd encouraged their rat habit with the Habitrail. Oh well, it was better than letting the rodents run loose in the trailer. If only they made a Habitrail for preteen boys.

Mark was a good guy. He might even be Mr. Right. But Judy wasn't looking for Mr. Anybody. She hadn't really ever gotten over her ex-husband, Mr. Have You Seen My Blue Shirt, The One With the Cocaine in the Pocket? His friends called him Clay. He called her to bring him bail money. Thank God that was over. And she'd probably never get over Juan. Fear, anger and guilt tend to stick with a person. Especially when you get called in to the morgue to identify the body before the ink on the restraining order is even dry.

"Where's Nan today?" Donna asked, bringing Judy back to the present.

"Walgreen's. Her meds were just about out," Judy replied, looking over at Nan's empty station, with hairbrushes and permanent rods strewn about like mobile homes after a tornado.

"Ooh. We wouldn't want that!" Donna rolled her eyes. Nan, who had worked at the station next to Judy's for almost three years, had multiple personalities. Not just voices in her head like most people. Official multiple personalities, each of whom had his or her own file at some therapist's office and would eventually make a fine movie on Lifetime. There were six of them – personalities, not therapists – at last count. Only two of them could do hair. The fun part was that no one was sure from one day to the next which ones would show up. It made life at Lone Star Cuts interesting for both the customers and the staff.

Except for Donna. *Poor Donna*, Judy thought. *I'd hate to have to reschedule Nan's clients all the time.* Donna always had eraser crumbs on her blouse, in her hair, and occasionally stuck between her teeth. Nan's clients would get scared by a bad appointment and cancel. But somehow, Nan convinced almost all of them to come back. And she kept bringing in new ones. Like magic. Judy thought maybe she recruited them from her group therapy sessions. Or prison. Who knew?

Donna rearranged the hair paraphernalia ("Haira-phernalia," Judy called them) on the front counter to while away the time. Purse-sized brushes. Travel sized conditioner. Hairspray bottles, arranged in order of size. *Big Hair, Bigger Hair*, and *Only in Texas Hair*. *Only in Texas Hair* sold the best, more than three to one over

Bigger Hair. Hell, you could get enough to hold up through two tornadoes and a sudden Texas downpour for $3.99, which was quite a deal, considering it was also the only hairspray that held up in the Austin humidity.

Judy watched Donna lazily, still thinking about the missing Nan. "We wouldn't want Nan to run out of her medication," Judy whispered, making her eyes bulge out and stabbing the air with her scissors. "That'd be too scary even for this place."

One Prozac or Two?

The sign might as well have said *Lone Star Cuts - Home for Dysfunctional Hairdressers*, because in truth, that's what the salon was.

April, who had owned Lone Star for a little over six years, had a great mind for business, but deep down, she was a patsy. It was fitting that Patricia was her middle name. It didn't matter who you were, if you came wandering into the salon with a sob story, she'd give you $5 and call you a cab. If you came wandering in with a sob story and a pair of scissors, she'd hire you.

To her credit, however, April had a knack for choosing depressed, mentally-challenged, emotionally unstable psychopaths who were usually pretty good at doing hair – with the exception of Nan, although 33.3% of the time, she was actually quite good.

Judy was the worst. Cutting, perming, dying, styling… She could even screw up washing hair, although that was her best event in the beauty shop pentathlon. But clients didn't seem to care. They loved her the way people love their five-year old granddaughter who leaves sticky handprints on the furniture and toys scattered across the floor. Judy was the reason business had picked up five and a half years ago, when she joined the staff.

Leon came with the place when April bought it from the past owner, a Chinese woman with a blunt cut, a blunter personality, and an odd smell. On some days, April swore she could smell whatever it was – and she didn't want to know – in the walls. Leon was a real piece of work. Straight male hair stylists were about as common in Texas as vegetarian rodeo riders. And Leon wasn't just straight. He was always horny and always full of good ol' Texas boy macho crap. April worried more about Leon than the rest of them, even though he'd never drawn blood. She could smell a sexual harassment suit in his future, in her future. But she just couldn't get rid of him. Where would he go? Besides, his mom was one of her best friends. That cinched it. Leon was there until Leon left, lawsuit or no lawsuit.

April had hired Ima a few months before Nan. Ima was psychic, which wasn't much of a problem, but she was also narcoleptic. Every day, April thanked her lucky stars that Ima did nails, not hair. She'd much rather have her fall asleep in the middle of an acrylic fill/palm reading than while shaving some client's neck.

Bernadette had only been there a year and was actually pretty normal. It was the guys she hung out with who were the problem. Sometimes they would show up in the middle of the day and hang out until Bernadette was done. One guy asked if he could park his motorcycle next to the shampoo station so he could keep an eye on it. Another one, Howard, April remembered him as, was 6'9" and over 350 pounds. He refused to sit down. He'd stand next to Bernadette's station for hours hulking over her clients. April was overjoyed when Bernadette dumped him. He scared her. She didn't care if he was a CPA in the State Treasurer's office.

Donna, the receptionist, was fine now, but she was a mess when she first started working at the salon. Her heavy metal star-wannabe boyfriend of three years had disappeared one afternoon leaving nothing behind but a chipped guitar pick and a holey AC/DC t-shirt that hadn't been washed in months. For weeks, Donna wore the t-shirt to work, still unwashed so she could be "close to his smell," and would break out into tears at the slightest provocation. That was over now and Donna seemed to have adjusted to her single life well, not to mention she'd regained most of her hearing.

There'd been others over the years. They came and went and kind of all blurred together. There was Elva, the self-styled female Elvis impersonator who was the best stylist to ever work for Lone Star Cuts. She'd left three years ago to seek her fortune in Las Vegas. April occasionally got postcards. Elva was living with two showgirls and learning to do make-up. Then she was dealing blackjack and living in a mobile home in the desert. April had lost track of Elva lately, but Elvis sightings at rural Nevada hair salons had increased dramatically.

And there was Jo-Jo, the only legally blind hairdresser in the state. Of course, April had hired her. How could she not? And, even though it had been a disaster, April hadn't fired her. Her husband got a job at a bank in Dallas, and she went with him. When she left, April prayed Jo-Jo wasn't planning to take turns driving the U-Haul.

April was friends with them all, even after they left Lone Star Cuts. She knew this violated one of the first rules she'd learned at the UT School of Business Administration: "Your employees are not your friends!" She could hear Professor Myron Smith's whiny little voice as he made this point over and over. And although she could understand the logic of it, April just couldn't help herself. She worked with these people day after day. She saw them more than she saw her husband, Bruce. If they weren't her friends, who would be?

She particularly liked Judy, partly because Judy always made her laugh. April tended to take life just a little too seriously, and it always helped to hang around someone who didn't. The fact that Judy could laugh at her awful life – the drugged ex-husband, the stalking former boyfriend who'd died in a drunk driving accident, the struggle of raising two boys alone – impressed April a lot.

Judy was rubbing off on her. These days, April laughed a lot more. She also cursed a lot more. (Everything had its unanticipated side effects.) It was only a matter of time until April would yell at Bruce, "Change the damn channel yourself!" He'd probably stroke out in his Barcalounger.

Meanwhile, April had a Christmas tree to have hauled away. Maybe tomorrow. After all, it wasn't even the new year yet.

Does This Year Make Me Look Fat?

The new year was only seven minutes away and Judy was looking forward to it like a D student looks forward to summer vacation. It may be delusional to think that marking off one day on a calendar could wipe away the past and start a life that is new and better, but delusion isn't such a bad thing when you've lived a life like Judy's.

This year, Judy thought, *I'm going to get in shape, learn math so I can help the boys with their homework, take a home repair class so I can fix the hold in the floor, and get up on stage if it kills me*. She paused for a second in her list of resolutions. *The one thing I'm not going to do is fall in love. Not again, not this soon.*

Of course the fact that she was making the last resolution while in the arms of Mark, her first serious relationship since Juan, slow-dancing to a middle-aged woman in a pink chiffon marshmallow of a dress who was belting out *Unchained Melody* in the ballroom of the Hilton was somewhat incongruous. Judy was either setting herself up for failure or success. It depended on your perspective.

Mark was a great guy, not bad to look at, and he did all the right things. This was, after all, the first time a date had actually taken Judy out on New Years' Eve to a party that required dressing up. Or even shoes. Judy had borrowed a green silk dress from Bernadette, because not only did she not own anything nice enough, even if she had, it would have been too big. Judy had lost thirty-three pounds since Halloween. No mean feat during the holidays. She hadn't raided the boys' Trick or Treat stash, made too much stuffing accidentally on purpose at Thanksgiving, or nibbled the arms off gingerbread men and claimed they were born that way. Still, the dress was tighter than Judy was used to, so she had spent what seemed like hours in front of her mirror at home making sure there wasn't any fat sticking out under her arms or on her back. She could just see herself twirling around on the dance floor, only to be hit by a couple of pounds of flesh making their escape from under the sleeves of the dress.

Judy had even toned down her makeup so she'd fit in better with the Hilton crowd. Not that there was anything wrong with the Hilton – she just wasn't accustomed to the lifestyle. Judy was more of a Motel 6 kind of girl. Motel 6 people were happy just to have an icemaker and that little paper strip across the toilet – they didn't need a thick, fluffy robe monogrammed with the name of the hotel and a bidet. Judy assumed the rooms at the Hilton had bidets. On the drive across town, Judy had reminded herself to watch her language and stand up straight tonight. She was used to having the rich and famous in her chair at work, but now that she was playing on their field, it was a whole new ballgame.

It feels nice here in his arms, Judy thought just as the song ended and the band kicked into a fast one.

Cut it out, she admonished herself. *That's how it happens. First you think, 'This feels nice,' then they move in and make your life a living hell.* Determined not to be seduced by the moment, Judy pulled away and started to walk off the dance floor. Mark accompanied her.

"Just one more?" he asked.

Judy shook her head. "I'm lousy at fast dancing," she apologized. "I look like I've just dropped pudding down my shirt and am trying to shake it out."

Judy could still feel the sting from the days when she was the fat girl at school whom all the kids, even the nerds, pointed at and laughed when she tried to boogey. Besides, she reminded herself, too much dancing led to marriage and children. At least that's what the Sunday School teacher at the Southern Baptist church she'd gone to in Abilene had always warned. Those were the last things Judy wanted now. She wanted to learn how to be alone. To not have to rely on a man. To face the world confident and self assured enough not to end up with a loser again just to be with someone. And she definitely didn't want any more kids. Two was more than she had bargained for. No matter how much she loved Stevie Ray and Nolan, the thought of another pair of feet pitter-pattering through the trailer was scarier than any of the horror movies she refused to let her sons watch.

Mark held her chair out for her. *He is a gentlemen, I'll give him that*, she thought as she sat down. "How sweet. Thanks," she told him.

"I'll go get us some champagne," he said as he headed toward the bar.

"Okay." Judy wasn't much of a champagne drinker. The bubbles made her light-headed, but it was New Year's Eve, and she was out of the house for a change.

While Mark was at the bar, she looked around the ballroom and saw that she really didn't stick out like the sore thumb she pictured herself as. A few tables over, there were four middle aged women celebrating like it was 1969. One, recently-divorced, Judy guessed, was wearing a skirt that could have doubled as a headband and knocking back Tequila shots. At the other end of the room, a twenty-something couple dressed in Goth clothes and party hats, looking like funeral directors at a Shiners' convention, danced without moving their feet. She relaxed a little.

Mark returned with two glasses of champagne and set them on the table. He pulled his chair closer to hers and stared into her eyes. "Can I tell you again how beautiful you look tonight?" he asked.

"You can." Judy squeezed his hand with a squeeze she hoped came across as friendly and thankful, not needy and lustful. She noticed again how soft his hand was. Juan's hands had been calloused from years of farm work. The first time she'd had sex with him, she had just about jumped out of bed when his rough hands practically sliced into her skin. She never really got accustomed to the feel of his touch and had insisted he use extra-moisturizing hand lotion every night, even though he complained that it made him smell like a sissy. At least that's what she thought he said.

"You're the prettiest girl in the room," Mark interrupted.

Judy looked around as if trying to determine just how big a lie he was telling her. She smiled despite herself. "It's not going to work, you know."

"Why, whatever do you mean?" he asked coyly.

"You know exactly what I mean. You can compliment me until the giant ball drops, but you're still not getting into my pants tonight."

Mark's face registered mock offense. "I'll have you know I have no intention of getting into your pants. You said you wanted to wait and I promised." He paused, and then grinned slyly. "Besides, I don't know if you've noticed, but you're wearing a dress."

Before she could say anything, the band started the countdown. "Ten, nine, eight, seven, six…"

Mark reached for her hand and pulled her up out of her chair.

"Five, four, three…"

He pulled her into him. He smelled good and the warmth of his brushed micro-fiber shirt felt silky against her face. It was the same shade of brown as his eyes.

"Two…"

He held her chin up. She was glad she wasn't wearing one of those silly hats.

"One."

He kissed her. His lips were moist and full. It wasn't one of those sloppy, "What does he think he is, a Wet-Vac?" kisses that so many men seemed to think was sexy.

There it was. THE NEW YEAR'S KISS.

This doesn't mean anything, she assured herself, pulling away slightly and looking out into the room. Because if the truth be known, that kiss didn't really do it for her. It was nice. And nice wasn't really Judy's cup of tea.

The Thighs Have It

It was so humid, Jud wouldn't have been surprised to see catfish swimming past her windshield. Judy's sweatpants were living up to their name already, even though the only exertion she'd had so far was pressing down on the accelerator. A slow but steady trickle of perspiration made its way down the back of her knees, past her calves and pooled into her socks.

"Why is it so damned hot in the middle of January?" Judy asked the radio DJ as if he were there in the car with her. She talked to him a lot, trying out jokes, polishing her stories, and complaining about life in general. It was a great relationship – she talked and he played music for her. Some days she thought that's really all she needed from a relationship, except for the sex. Although she had a pretty good handle on that herself these days.

The DJ had no answer for her and his voice was staticky as he announced the next song, so Judy pounded on her dashboard. It was a trick she'd picked up from her ex, Clay. He thought everything would improve if you hit it. That was one of the reasons she'd divorced him. That and the fact that he had no sense of humor. Not to mention his pesky habit of calling her at 2 a.m. from the Austin Police Department.

But the radio continued to buzz like a bee trapped in an open jar of peanut butter left out all night on the patio, and now her hand hurt. One of these days, when the boys were grown and she was a famous stand-up comic, Judy was going to get a new car. One with a working radio. Working brakes would be nice too.

Judy turned off the radio as she pulled her car, a 14-year-old beat-up Dijon-mustard yellow Subaru station wagon she'd bought used right after her divorce four years ago, into the Zilker Park parking lot. The brakes squealed, and as a bonus, when she switched off the ignition, the whole car shuddered as if a ghost had just walked over its grave. The rear bumper – which sported a bumper sticker that said, "Practice Senseless Acts of Beauty, See Your Beautician" – was barely clinging to the chassis. Leon had helped her put the bumper back on with half a roll of twine the day she had backed into the dumpster at Lone Star. The bumper sticker was a gift from her best friend Lynn, who had had it printed up for her at the mall. The same friend who was forcing her to be upright and sweaty this morning.

Nobody here, Judy thought, hoping it meant maybe she could go home and go back to bed. After all, it was her day off. She should be home sleeping in her four-poster bed with 800-count Egyptian cotton sheets as her perfect husband fixed her croissants and fresh raspberries while her two perfect children picked flowers for her from the rose garden kept immaculate by her hunky gardener.

Yeah, right. In another life maybe, if she was lucky and landed a role on *Desperate Housewives*. In this life, Judy lived in a run-down trailer park where the closest thing to a rose garden was the jasmine-scented deodorizer stuck to the laundry basket to cut down on the smell of dirt, sweat socks, and rat poop. Her boys spent most of their weekend mornings arguing at full volume over whether they were going to watch cartoons (Nolan) or play videogames (Stevie Ray). And while Mark might be the best guy she'd dated in a long time, maybe forever, she was already debating whether and how to dump him without hurting his feelings. She'd managed not to have sex with him yet and he'd been pretty patient about it, but even so, she could tell the relationship wasn't going to work. He was a happy, honest, hard-working guy who seemed to adore her. Not Judy's type at all. He had no dark side and Judy had always been drawn to the dark side. Donna called it 'Luke Skywalker Syndrome.' *I'll have to dump him before Valentine's Day because I don't want him buying me jewelry*, she thought. *It'd be just like him to go and do something unreasonable like that.*

Judy drank the last few sips of the only cup of coffee she was allowed for the day from a beat-up plastic travel mug. The coffee was tepid and watery, but caffeine is caffeine. She'd been addicted to coffee since she was twelve, when she would sneak sips out of her dad's mug every morning before he'd leave for his job as manager of the bowling alley. But as far as addictions go, coffee wasn't so bad. That and Midol were her only real drugs, and she could live without the Midol – although everyone else would be miserable.

As she opened the car door, she accidentally dislodged a Spongebob Squarepants toy that had gotten wedged between the driver's seat and the door. Too bad Spongebob had fallen out of favor with her boys. He was so harmless, not like that stoner, Scooby Do. She still had that old Spongebob Halloween costume of Stevie Ray's. Nolan would probably never wear it. He couldn't be seen in cartoon clothing now that he was a "Big Boy." There were days he refused to wear his Spiderman underwear because "Spiderman's not cool, Mom!" *Tell that to Tobey Maguire,* she thought.

Looking around the parking lot, she spotted Lynn standing next to a tree, staring impatiently at her watch. Looking great, as usual. Judy glanced down at her own paint-encrusted sweat pants and a t-shirt with the words

"Runs with Scissors" emblazoned across the chest. Everything Lynn wore screamed out "ensemble." Everything Judy wore screamed "Garage Sale Free Box!"

Judy tossed the travel cup and Spongebob back in the car and locked the doors. She reached up under her t-shirt and stuck the key in her bra. The boys had given her one of those little key-holders that you attach to your shoe for her birthday after she had declared her intention to lose weight, but she'd worn the Velcro out before she'd ever had a real reason to use it. She couldn't find it now. It was probably in the kitchen junk drawer with the dried up super glue, the grocery receipts, birthday candles, batteries with no charge that she couldn't bring herself to throw out because they were bad for the environment, coupons, jokes written on napkins and envelopes, plastic pieces from toys that had long been thrown out, rat droppings…

You can tell a lot about a person by what they had in their kitchen junk drawer. In Judy's case, it was pretty much the same stuff she had in her head.

"About time you showed up," Lynn yelled across the parking lot in that voice she used that could be teasing, but probably wasn't. Lynn considered five minutes early to be ten minutes late, a luxury childless unmarried women have without realizing how lucky they are.

Judy sauntered over slowly. "Hey, I'm here." She yawned, more for effect than out of necessity. "I don't see why we have to do this before dawn."

"It's 8:23. See that?" Lynn pointed to the horizon. "Sun's been up for hours." She perused Judy's get-up, peering over the top of her Foster Grants. "Nice shirt." Judy didn't know if she was being facetious or sincere. She went with sincere. Why not?

"Thanks! April gave it to me for my birthday." Judy's friends knew that the key to her heart was through her funny bone.

"Please let me take you shopping next weekend. With all the weight you've lost, you need to start showing off your body." Lynn smiled. "The guys won't be able to take their eyes off you. The women either." Lynn stared at her with mock lust.

They both laughed.

"We'd better get going," Lynn said as she moved out from under the tree and started walking on the newly-replenished bark-o-mulch trail. It was Lynn's warm-up pace, but to Judy it was more like speed walking.

"Slow down, Speedy Gonzalez!" Judy coughed. "You got someplace you gotta be? A late-breaking story you have to get right on?"

Lynn shaved a few m.p.h. off her pace. "Sorry. I forgot. You're thirty-three now. You can't move as fast as you used to when you were young." Judy's birthday was the week before Christmas. She and Mark had only been dating eleven days at the time, but he had still gotten her a present – tickets to Faith Hill and Tim McGraw. The tickets were so expensive, Judy insisted they counted for Christmas too. Judy had loved the concert, except for the fact that Mark had kept his arm around her the whole time. She thought he was acting too possessive and clingy. She wasn't used to all the attention. Clay and Juan had often acted like she wasn't there at all.

"How does it feel to be middle-aged?" Lynn asked taunting her.

"I am NOT middle-aged," Judy argued. "Everyone in my family lives to be a hundred or more." *Except mom*, Judy thought. Her mother had died of breast cancer when Judy was fifteen. If Judy had been the type of person to blame her own miserable life on her tragic childhood and had the money to afford counseling, a therapist would have been able to afford few extra Lexuses – or was that Lexi? But Judy just kept moving forward, putting her past behind her and trying to do the best she could with what she had. And when she had a chance, she watched Dr. Phil.

"I won't be middle-aged for another seventeen years, thank you," Judy said emphatically. She looked over at her friend, whose arms were now starting to pump. "On the other hand, you're thirty-seven. That's just three years away from the big 4-0."

"Thanks for reminding me. I swear if you do that whole *Over the Hill* crap for my fortieth, I'll never talk to you again!"

They walked in silence for a few seconds, with Judy trying valiantly to keep up the pace. Lynn looked over in admiration at her friend. "You're much faster than just a couple weeks ago. And the weight's really coming off now. You look great."

"You only think I look great because you can't see what I'm hiding under these sweat pants. My thighs rub together so much, I had to spend extra money for fireproof pants." Judy never could take a compliment, probably because she hadn't had that many.

"Jeez, Judy! I try to say something nice and you…"

"Sorry. Thanks." Judy looked down at her thighs. They were definitely thinner and stronger than they'd been in years. Not Catherine Zeta Jones' thighs yet, but she was getting there.

"You look great. I told you working out was the way to go, didn't I?" Lynn cocked her head. "It is our workout, right? Or is it sex with Mark that's toning you up?" Lynn teased.

"We haven't had sex." Judy reminded her.

"Still?" Lynn was surprised. "I'm glad you're taking it slow. Mark's a great guy. I wouldn't have set you up with him if I didn't think you two would be perfect together."

Judy didn't want to burst her friend's matchmaking dream right then by telling her she planned to break up with the great guy as soon as possible. "Yes. Slow is good. I like slow."

As if to be ironic, Lynn sped up the pace. Judy had to flail her arms like a windmill to keep up, but when she noticed a squirrel sitting on the grass next to the path, she pulled over and bent down. As much as Judy was frightened by rats, she loved squirrels.

"Can you believe she doesn't even realize I'm not there? You'd think the fact that I'm quiet would be a dead giveaway." Judy sucked in a big gulp of thick wet air. "This is easy for her. She's like you. Always out running around getting exercise. Gathering nuts. Doing what she's supposed to be doing. Me, I used to spend my mornings with my face in a box of donuts. You wouldn't have one stashed away around here somewhere, would you?"

The squirrel didn't answer. He kept doing whatever it is squirrels do when they just sit there on their hind ends looking cute. Lynn reappeared behind Judy.

"Wanna finish the walk or are you and the squirrel trying out a new comedy routine?"

"He says he works alone. Damned shame. I think we'd make a fine duo. Like Burns and Allen." Judy loved old comedy acts. She had a boxed set of DVDs featuring famous comedians of the 40s, 50s, and 60s.

"Or Lucy and Ricky. She never let him get a word in edgewise either."

Judy slapped at Lynn's arm but missed. Lynn had already started moving again. She didn't want to fall below her target heart rate for even a nanosecond. Judy jogged to catch up with her. The short rest had done her good.

"Okay, we've talked about my new body, my lack of a sex life, and my lack of a comedy career. Let's talk about you for a while. What's up in your world lately?"

"Just work," Lynn sighed. With Lynn it was always about work. She worked for the *Travis Sentinel*, Austin's alternative newspaper. In the newspaper world, they were definitely number two. Lynn had once joked that their motto should be "We get our news from the other newspaper too."

"Are they still laying people off?" Judy asked.

"You mean 'right-sizing.'" Lynn grimaced. She hated it when the people at the top thought renaming something made it hurt less. She'd rather just have someone come right up to her and tell her "You're fired," than have them pussyfoot around and avoid looking her in the eyes. Where was Donald Trump when you needed him?

"Right-sizing, huh? Sounds like picking out a bra," Judy joked.

Lynn didn't laugh. " They 'right-sized' two people last week. Katy in the mailroom and Don in classifieds."

"Don? Isn't he the one with the bad hairpiece who keeps asking you out?"

"Yeah, that's him."

"Well, I'd think that would be good riddance. The man was clueless. Not to mention greasy."

"Yeah, I'm not sorry he's gone. He can't accept that I'm, in his words, 'committed to the lesbo lifestyle.' He's sure I'd get over it if I'd just do it with him. What a jerk!"

"It's just self-defense," Judy said. "A penis is the only thing they have that women want. Some women. So, when you tell them to take their penis for a hike, they get defensive." *Or they get plastered and run their car into a concrete highway divider,* Judy thought.

Lynn had her own demons to fight. These days, when she walked into her office at the paper, she swore she could hear her blood pressure rise. Forget ringing in the ears, she had whooshing, like someone had turned on a fire hydrant and forgot to turn it back off. Even if she didn't get right-sized, she would probably stroke out at her desk. The cleaning lady, Yolanda, would probably come by late at night and find her lying prone on the floor, the computer keyboard still firmly in her grasp, ready to finish her assignment before the deadline, rigor mortis be damned. "People keep disappearing. It's like a rerun of *X-Files*. Very scary."

"You want scary? Try working next to Nan. You never know who'll show up. It could be Harriet, the Martha Stewart wannabe or Robin, the angry transvestite." Judy had a way of talking about me-me-me, but this time she did it on purpose, to get Lynn's mind off her troubles.

"Yeah, but at least you never have to worry about losing your job. April would never fire anyone. It'd kill her."

"True." Judy's shoelace, which had been loose since she left the car, came completely undone. When she bent down to retie it, Lynn actually stopped and waited. "But it can have its downside, you know. April could hire a serial killer as a shampoo girl tomorrow and we'd be stuck with her, shuffling our customers around the chalk outline on the floor, 'til the cops showed up to drag her off in handcuffs." Judy stood back up. "Besides, I don't know what you're worried about. Your boss loves you. You just got that big byline last month. I have the article up on my refrigerator – 'Recycling – Environmental Deja Vu.' See, I even remember the title."'

Lynn started walking again. "Hell, my boss could be the next one to go. It figures. I finally get a good job at an actual newspaper and the whole world decides it wants its news from the Internet."

"Not me," Judy huffed. "I like to have my newspaper right there with my bowl of Cheerios. I figure I burn off my whole breakfast getting pissed off at the idiot stuff that happens in this town." Judy loved reading the newspaper every morning. Knowing what was going on in the world made her feel smart. It wasn't something she felt very often. Not since she'd dropped out of high school to help her dad run the bowling alley and help take care of her sisters when her mom had died. "Besides, I've tried clipping coupons off the computer screen and I keep breaking my scissors." Judy grinned, trying to egg Lynn into a smile, a smirk even. Anything.

"Yeah, that's the only reason people read the paper these days. For the coupons. One of these days, I'm going to be stuck writing coupon copy." That was Lynn. While some people see the glass half full and some see it half empty, she always saw the glass sitting on the edge of the counter about to topple off.

"You could do that. 'Two-for-one Depends. For the career gal on the go. In all the latest fashion colors...'" Judy looked over at Lynn to see if there was even a hint of a smile yet. None. Dammit. It was going to be that kind of walk. No laughs. Lots of whining. So Judy changed subject. "You should let me do something with your hair."

"My hair is fine."

Judy kept pestering. "You'd look so cute with a short cut."

"Cute, yeah, that's what I'm going for. Cute will save my job." Lynn was in some mood today. "Besides, I've had short hair. I looked like a dyke."

Judy couldn't help but laugh. "And that's a bad thing?" Lynn was her only close lesbian friend, and Judy still wasn't sure how the whole thing worked. Not the sex part. She got that. The relationship part. Judy had once suggested that Lynn write an article called "A Lesbian Starter Kit." What to wear, who leads when you dance, who calls whom. Lynn had ignored her because she wasn't "out" at work. Except to the jerky guys who kept pestering her. No one believed them anyway. All the women claimed to be lesbians around them.

A few minutes passed, with only the sound of their sneakers crunching the bark underfoot. Eventually the crunching was interrupted by footfalls of another morning exerciser, coming up behind them. As he passed them, they both got a good look at him from the back side – a good-looking guy with black shorts, a tank top, and no back hair. He pulled away quickly, not even saying hello.

Judy stared after him appreciatively. "Nice ass!"

"Why don't you jog after him?" Lynn teased.

"I'll just admire the scenery from afar."

Lynn's train veered back onto its one track. "I think it'd kill me if I had to stand in line to get food stamps and unemployment benefits."

"Well at least you've got those $280 Jimmy Choos so you'll look good there. Quit worrying."

Two women runners came around the bend, wearing sports bras and short-shorts.

"What's next? Thong leotards?" Judy lamented.

"Huh?" Lynn asked, obviously distracted.

"Oh, I see. If I'd known that was all it took to get you out of your blue funk, I'd have worn short shorts too. But never a thong."

Lynn laughed fully. Finally. She said something, but Judy couldn't hear her because a car drove by with the windows open and the radio blaring so loudly, Judy could feel it in her cellulite. Lynn scowled at the teenagers in the car.

“What did you say?” Judy asked after the car got far enough away for her temporary deafness to subside.

"I said she, that runner, had a lot of muscle."

"Noticed that, did you?"

"Yeah. And muscles weigh more than fat."

"That a fact, college girl?”

Lynn raised the right half of her lip, Elvis-like. She didn't like it when Judy called her 'College Girl.'

"Yep. So don't be surprised if you gain some weight as you get in shape."

"Me in shape? Now that's a concept!” Judy sighed, "I just wish all my weight wasn’t coming off my chest. Pretty soon I’m going to have to wear a bra with inverted cups. I wonder if I can get one at Victoria’s Secret.”

Judy took another mental detour and again, it was not to a happy place. "By the way, did I mention my gynecologist wants me to get another mammogram? She says it’s nothing to worry about. Just my regular lumpy breasts.”

"Couldn’t hurt," Lynn offered weakly.

Judy stopped walking. "Spoken like someone who’s never had a mammogram.”

Lynn ignored her. The truth was hospitals, labs, x-ray machines, they all scared her. As a born pessimist, being around people who were sick and dying always made her think the worst. So she avoided doctors and hospitals as much as possible. She hadn’t had a pap smear in six years. Judy didn’t know that. She’d have jumped all over Lynn if she did. "I’m sure everything will be okay," Lynn chimed in, with her most positive voice.

"Of course it will," Judy said convincingly. "A few bruises, a hickey on my nipple, and I'll be just fine.“

They both laughed. "I just want to get it over with. My appointment isn't for three weeks, though. You know, you could come with me and get one too. We could make it an adventure."

"I'll pass," Lynn said like she'd been offered scoop of goose shit ice cream.

Good versus Bad

After six years of putting up with his boozing and collecting DUI tickets like they were Hot Wheels, of watching him come home and hit things (although to his credit, he never hit her or the boys), Judy had finally thrown Clay out. She couldn't even remember what the final straw was, or if there even was one. She just called the locksmith one day and had the locks changed. When Clay came home, his key didn't work and all his clothes – most of which smelled like booze and cigarettes – were lying on the lawn. He'd pounded on the door for a few hours, just long enough to get sober enough to decide to hit the bar again or go see his dealer, she wasn't sure which. She was just glad the pounding had stopped. So were the neighbors.

He had called a few times after that night and made promises, but she'd heard them all before. She told him they were through and he needed to accept it and move on, and then she demanded full custody of the boys. He didn't really fight her, which made her sad in a way, but it really was for the best.

On most days, she considered herself pretty lucky to have gotten out of the marriage with no desire whatsoever to join a group of women who sat around every night exchanging recipes that included boiling their exes testicles in oil. As Judy reminded herself whenever she was feeling especially down, if she hadn't married Clay when she was twenty-two and hadn't spent three years working at Dairy Queen in a hair net trying to keep up with rent and food and bail money, she might never have gotten the motivation to go to beauty school. And then where would she be?

She'd probably still be 5'5, 210 pounds, living at home with her alcoholic dad, helping him out in the bowling alley, doing his laundry, and trying to not make eye contact with the long line of floozies padding around the kitchen in the morning looking for a drink. And she'd still be in Abilene, Texas, a place whose motto should have been "It ain't the heat, it's the stupidity." She'd gained thirty I.Q. points just moving away.

So, yeah, Clay was a mistake. And yeah, he was an asshole. But Clay had been her ticket out, and for that she silently thanked him.

Juan was mistake #2. Even when she was racked with guilt for having thrown him out on the day he ended up splattered across the freeway, she knew that it was the right thing to have done. And that she really wasn't to blame for the accident. Although some days, especially when her blood-caffeine-level (BCL) was too low, she wasn't so sure.

The good news was, Judy was done with them both. No looking back and wondering "What if?" Judy wasn't the kind of person who looked back. The only time she ever used her rear view mirror was to check her mascara or make sure the boys hadn't killed each other in the back seat.

If she was going to spend her time thinking about the men in her life, she definitely should be thinking about Mark, not the bad boys of her past. But therein was the problem. Judy liked bad boys, at least at the beginning of a relationship. The adrenaline rush made her feel sexy and dangerous, not like the frumpy mom she felt like most of the time. Her whole body would buzz and her heart would feel like it was going to burst right out of her chest. Later, when the bad boys ripped her heart out and ran it over with their Harleys, then backed up and did it again, she admonished herself to never DO THAT again.

Yet somehow, even knowing the pain that inevitably lay ahead, Judy was still hooked on the rush of men who she knew would abuse her. Mark didn't give her that rush at all. That's why she was going to break it off with him. If only she could find a way.

My Trailer Park or Yours?

Judy wandered barefoot onto her front stoop, carefully avoiding Stevie Ray's skateboard and Nolan's collection of rocks. She found the Sunday paper standing upright, wedged between the rusty legs of her porch chair and the slightly bowed-out side of her trailer. At least this time the paper boy had managed to get it close. Although Judy couldn't complain since she wasn't paying for the service. The woman one trailer over had a two-year subscription when she moved away and Judy had inherited it. Lynn didn't like it that Judy read the "other" paper too. She felt like it was some kind of betrayal, even though Judy didn't pay for it or believe half of what she read.

The morning air was muggy, but chilly. It felt like taking a bath in someone's used bathwater. A feeling Judy knew well, because when the boys were little that's exactly what she did every night to save water.

A cat mewled from somewhere close by. *Probably another feral cat about to have kittens*, Judy thought. Last March, Al, the mobile home park's kindly but inept manager, attempted to round up all the feral cats in the area and get them neutered, but there were just too many of them and they were just too fast. *I'd be fast too if I thought someone was going to snip at my private parts*, Judy thought. She listened closely to hear if the cat sounded like it was hungry or in distress, but before it made another sound, three teenage boys ran down the circular drive that interconnected all the trailers in her section of the park. Each had one hand on the waist of his baggy pants, trying to keep them up to avoid tripping. *Well, at least they can't steal anything big, they only have one free hand*, Judy thought. She hated those baggy pants. Stevie Ray had four pairs. She tried not to look at him below the shoulders when he wore them. The good news was, they were so big, he could probably still wear them well into his forties.

Newspaper in hand, Judy tiptoed back inside the trailer. It was peaceful, a treat as rare as her child support check from Clay arriving on time. But then, it was only 7:40 a.m. on a Sunday. Judy usually didn't usually get up until 9-ish and then only because the boys had the TV blaring so loudly there was no point in trying to go back to sleep. It hadn't been so bad before they got cable, because the boys had had to choose between a home improvement show with lousy reception or The Reverend Billy Ray Bennett and His Hour of Gospel Power, which came in loud and clear. Now there were twenty-eight channels. The boys spent most of their time clicking back and forth. To Judy, lying in bed trying to catch a few more minutes of sleep, it sounded like one fast-paced gardening-QVC-cartoon-right-wing-newscast-basketball-soccer-religious-Mexican soap opera.

She'd tried stuffing cotton balls in her ears and even wearing a pair of ear plugs one of Bernadette's ex-boyfriends, a target shooter with a missing ring finger, had given her. But just knowing the boys were up was enough to keep Judy's mind from letting her fall back asleep. It was on red alert, or at least fuchsia, whenever they could possibly be getting into trouble, which was every waking hour and plenty of the sleeping ones as well.

These days though, after having spent the last few weekends getting up early to go walking with Lynn, Judy woke up before her boys. But Lynn had called the night before coughing and sneezing, and had cancelled on her. Judy wasn't about to go walking by herself. She didn't have the motivation. Besides, it was nice to have a day off from everything except being a mom.

Abbott, one of the two stray cats Judy had invited in last year after Al's failed catch-and-release program, crossed her path as she returned to the kitchen, paper in hand. He cried loudly, which woke Costello from her perch atop the refrigerator. She jumped down and joined her feline co-conspirator in begging for an early morning treat. Judy gave in. She put the paper down, retrieved a box of Friskies and quietly shook a few into their bowl on the floor. The cats pounced on the unexpected bounty, pleased with their good fortune.

Content that the feline members of the family were satisfied for now, Judy sat down at the kitchen table and opened the paper. As she scanned the headlines, a breeze found its way inside from around the poorly-insulated kitchen window. The air was chilly and smelled like salt, as if during the middle of the night, Austin had been transported to Galveston, which wouldn't have been all that bad because Judy had always wanted to live near water. The "River View Trailer Park," which had neither a river nor a view, was the closest she'd gotten so far. It had been her home for the past seven years. At only $435 a month, it would probably be home for the rest of her life, unless one of her boys became a business tycoon and decided to buy her a fancy house on Lake Travis. Or she became a famous stand-up comic. But neither of those was going to happen any time soon, so Judy decided to pray for a tornado to pick up her place and move it to the coast. It was the only way she could afford beachfront property.

The coffee maker made that fat guy with sleep apnea noise it makes when the coffee's ready, so Judy got up to pour herself the one cup she was allowing herself today. Two packets of Splenda and a little non-fat milk and it was perfect. She sat back down and separated the newspaper into sections:

- Coupons to clip out later. Judy cooked by coupon; she found it easier than using recipes. Whatever could be had for 50 cents off, that's what they had for breakfast, lunch, and dinner. She also cleaned by coupon. Which would explain why the windows were always decently clean, but the toilet only rarely so.
- Funnies to read first. Judy's philosophy was that you'd better start the day with a laugh so you could get through the rest of it without getting grumpier and grumpier.
- Sports/classifieds/real estate section to put in the recycling bin without a glance. She knew that one day soon, she'd probably have to morph into a sports mom, but so far Nolan was satisfied with bike riding and Stevie Ray was never going to become a professional skateboarder despite the fact that it was the only non-video activity he was interested in. Until either one of her boys took up soccer or baseball or synchronized swimming or whatever, she felt no need to read anything about sports. And since she couldn't afford anything, the classified and real estate section was useless to her too.
- Everything else. This was the good stuff. Politics. Crime. Social events. Where the real laughs were. That's why Judy liked to save it for last. This section of the paper gave her material.

Judy read the funnies. First *Peanuts.* Old stuff, but she thought Charles Schultz's repeats were better than most of the other cartoons. Snoopy was kicking his dog food bowl up the front stairs to the Charlie Brown's door. The boys begged for a dog almost as often as they begged for all the other things they didn't have that everyone else had, but

Al didn't allow them. You could have forty-seven cats, but dogs were verboten.

On to *Beetle Bailey.* Judy read it every day even though she almost never found it funny. Force of habit. Sometimes she wondered who laughed at Beetle getting beat up by Sarge all the time.

Finally, *Stone Soup*, her favorite. Alix and Holly were fighting over whose turn it was to do dishes. Just like Stevie Ray and Nolan. Judy had to laugh. Especially at how outraged her boys would be to know two cartoon girls reminded her of them.

Finished with the comics, Judy started in on the front page. Crime and politics. Or were they the same thing? "Governor's Popularity at All-Time Low" shouted one headline in a font that was much too bold for early morning viewing. "Former Bank Employee Charges Republican Candidate Dick Sloan with Sexual Harassment," said another. Business as usual.

She read the Sloan article. "According to a spokesman for Mr. Sloan, the alleged charges are completely false."

Alleged. Judy liked that word. It would come in handy in her own life, she predicted. "So, Principal Smith, what is it that my alleged son Nolan allegedly did today?"

Politics could be my fallback career, if stand-up doesn't work out, Judy thought, smiling. *They're practically the same, except I could funnel campaign funds into Nolan's college fund.* There wouldn't be a college fund for Stevie Ray. Judy already knew that he was going to end up laying blacktop or installing cable for a living. Sending him to college made about as much sense as sending Howard Stern to charm school. Or Karl Rove to ethics classes. There was still a chance for Nolan, though. He did okay in school and actually seemed to enjoy subjects like math and science. Where that came from, Judy had no idea. But it was a good thing, since

Nolan was klutzy and obviously would never make it if he had to work with his hands.

I could get campaign volunteers to clean up around the trailer, she thought. *Maybe date a Secret Service agent.* "Does the Governor have Secret Service agents?" she wondered aloud and then shot a quick look toward the boys' room. Not a peep.

A twinge in her right calf caused her right leg to kick involuntarily. Judy got up and leaned forward on her chair, stretching the offending muscle in a standing runner's pose. Three months ago, she wouldn't have known what a runner's pose was, but now that Lynn had sucked her into the world of the active, she had learned all kinds of new things. That muscle weighs more than fat. That her target heart rate was one hundred and something. That a pedometer wasn't something you wore between your toes after a pedicure…

The stretch wasn't helping much, so Judy strolled around her tiny kitchen. Finally, the pain subsided and she sat back down to read. Page two. A computer glitch had screwed up grades for the entire freshman class at The University of Texas. Cows moseying down Manor Road had caused a three car pile-up. A bank robber dressed in an Eagle Scout uniform had held up Trans-Pacific Bank during lunch hour. *I wonder if he was working on his survival badge,* she thought with a smile.

Finally. The funny stuff.

Suddenly she heard Stevie Ray's voice from the bedroom. "Mom, can we go to the mall? I told the guys I'd meet them there."

Judy sighed. Real life interrupted once again. Funny how it always ended up doing that.

"The mall isn't open until ten. We can go then. Come get some breakfast." Judy thought for a moment about getting up and making a hot breakfast. Pancakes, bacon, some eggs for protein. That's what good moms did for their kids on the weekend. At least in the stories she read in women's magazines. But if she as much as pulled out a frying pan or cracked open an egg, her boys would think she'd finally lost all her marbles. Besides, they liked Count Chocula. Who didn't?

Nolan shuffled in, sleepy-eyed.

"I thought you'd be at the park exercising." Wow, even the boys thought of her as an exerciser. Well, it was better than being The Fat Mom. She'd been that for most of their lives and was plenty willing to let some other mother take the crown from her. Although sometimes, like when she made brownies from a box for Nolan's class, it was really hard not to cave in. *What's so bad about being The Fat Mom again?* she'd wonder. But lately just looking at her quickly-changing body was all it took to give her the willpower not to eat a whole bowlful of the chocolaty batter and wash it down with a half gallon of whole milk.

"Nope, not today. Lynn's sick." Judy ran her fingers through her younger son's messy hair, but he quickly slapped her hand away. He was at that age where he needed to prove his manliness at all times, which would explain why suddenly the bathroom was messier than it used to be. "Why do you want me out of the house?"

"No reason." He looked guilty, but most boys his age usually do.

"What kind of trouble were you planning to get into?" Judy could read him like the comic books he left lying all over his bedroom floor.

"Nothing. Jeez, mom."

Stevie Ray came in with his pants hanging down three inches below his underwear. At least his tidies were white. Judy had done laundry last night. She'd had to. She'd told Mark she couldn't go out on Saturday because she had to catch up on chores at home and spend some time with her boys. Doing laundry and watching DVDs with Stevie Ray and Nolan helped her prove to herself that it wasn't really a lie.

As convinced as she was that Mark wasn't right for her, or she wasn't right for him, or maybe the timing was all wrong, Judy hadn't been able to bring herself to break up with him directly. He was just too sweet and she didn't want to hurt him. She was used to being hurt, so she knew what that felt like. Lately she had started slowly pulling away from him, hoping he'd break up with her. He hadn't shown any signs of getting the message yet, but it was only a matter of time.

Stevie Ray steered her train of thought off its tracks, stomping his Yeti-sized feet. "Mom! What about the mall? Can you drive me?"

Judy nodded. As much as she didn't like being his chauffeur, she knew she'd like it even less when he got his driver's license. Too bad she'd told him his real age. It was only four and a half years until he'd be out on the road endangering his life and everyone else's. If she'd only lied and told him he was two when he was five, she could have bought herself – and the entire Travis County metropolitan area – some time.

Stand By for an Impotent Message

Ima sat, mesmerized by story on *All My Children* despite the fact that she usually knew what was going to happen before it played out on the screen. That was the worst part about being psychic. Always knowing the solution to the mystery, the punch line to the joke, how the relationship would end, beforehand.

Leroy walked by, spotted Susan Lucci on the screen and stopped dead in his tracks. He had a thing for good looking women, and unlike most men, he thought women over fifty, sixty, seventy were just as exciting as those under thirty. They were definitely more grateful.

Leroy's client, who had been following closely behind him, her hair in a towel, ran smack into his backside as he stood with his eyes glued to the tube. "Sorry," she said, more from force of habit than sincerity. If you are over fifty, female, and from the South, "Sorry" is one of the first words you learn as a child. That and "Yes." As in "Sorry I could only take fifty children on the field trip and not the whole class" and "Yes, I'll marry you even though I know it'd be less painful to set fire to my eyebrows." Old habits were hard to break. And just as hard to live through.

A commercial for dishwashing detergent came on and Leroy's attention turned back to his client, who had backed up slightly and was waiting patiently. "Come on, girl, we've got to make you beautiful!" You could hear from his tone he believed it was possible. That's what Leroy's clients liked most about him, the ever-present hope that with a little bit of spray and mousse, they could regain the beauty that had packed its bags and moved away decades ago.

Judy's station was far enough from the TV that even at full blare, she could barely hear it, which was okay because she wasn't really much of a soap fan anyway. Life was melodramatic enough. Besides, she didn't like the competition for attention. If Judy had her way, she'd be the only entertainment in the salon. She stood perched, one foot propped up against the other knee like a giant number 4, chewing gum and scrunching up her nose with indecision.

"I'm just reviewing the multitude of options," Judy finally said. "There's the spiky bang, the long bang, the short bang, the really short Betty Page bang, the asymmetrical bang. You can't just rush something like this."

"Just a trim," Lynn implored. "I really don't have all day. I have to get back to work to see if there's a pink slip on my door."

"You don't have a door," Judy said while popping a stick of Doublemint, an on-again off-again habit since she'd given up smoking three years ago, when she was afraid her boys would do as she did, not as she said.

The lack of a door was a sore point with Lynn. She'd always imagined herself as the kind of journalist worthy of her own office. But it hadn't happened yet and she was almostfortyforgawd'ssake (it was one word when Lynn said it, and she had considered herself almostforty forgawd'ssake since the day she turned thirty-five). If she got pink-slipped now, chances of a nice corner office with her own secretary were as likely as the chances of stumping for a right wing neo-conservative who wanted to ban gay marriage. “You’re right. I should stick to the facts. I’ll probably have a pink slip attached to my cubicle wall. With a pink thumbtack."

“You know, we could dye your bangs pink in a kind of a ‘Ward-off-the-pink-slip’ hair voodoo ritual,” Judy said, taunting her friend.

Lynn rolled her eyes.

Judy grabbed a water bottle. "Okay, be a stick in the mud. Let’s wet these bangs down and whack ‘em off then.” She spritzed. The moisture filled the air and mixed with a whiff of scented marker that was coming from the front desk. *Raspberry*, Judy thought sniffing, although Donna could have told her it was cherry. Donna was using the marker to update the specials board. In big letters, she changed “Acrylic Nails and Palm Reading by Ima - $45” to "Acrylic Nails and Palm Readings by Ima - $40 This Week Only!” Sure nails were only $30 a couple of blocks away, but when it comes to the price of beauty, it’s all about location, location, location. And Lone Star Cuts was the closest salon to the state capitol.

Suddenly, from the chair next to Judy’s came a polite but firm yelp. Judy looked away from Lynn’s head over to Nan’s station to see what the trouble was. Sarah Sloan grimaced as a Nan held a handful of her bright red hair too firmly in her bony hand.

“Nan!” Sarah chided with a refined Texas twang, “Remember, the hair gets teased, and the scalp prefers to lie flat against my head.” Sarah was in her late forties, but looked younger, especially today, dressed in a peach silk suit and large gold bracelets and earrings. The ensemble said “MONEY!” louder than the gaudy diamond rings many of the salon’s older clients wore to announce their social standing in this town where cold hard cash and the things it could buy were the primary language. Sarah was one of the richest women in the state but felt no need to beat anyone over the head with the fact. Maybe because she'd made her money herself.

How a powerful, politically-connected woman like Sarah Sloan ended up choosing Nan as her stylist was a mystery to almost everyone. Judy thought it must be part of Sarah's charity work, or maybe she'd lost a bet at a high stakes Texas Hold 'Em poker game.

Nan put down her rat-tail comb. "Sorry," she said not sounding as if she meant it. "Big enough?" Nan had an anemic look, especially for a Texas girl. A spray-on tan still left her with the complexion of a panty liner, which coincidentally, was about how wide she was. Standing next to Nan in front of mirrors all day had never helped Judy's self esteem.

"Any bigger and you're gonna have to check it with your luggage on the flight to Dallas," Judy joked, looking at Sarah, making eye contact as if to say, "Let me do your hair. I'm not all that talented, but at least I don't switch personalities mid-snip." With Nan as your stylist, you definitely took your chances.

"We're taking the private jet, Judy. We don't check luggage," Sarah laughed, her words not uppity or condescending, just a statement of fact.

Judy perked up when Sarah called her by name. It never hurt to have someone as important as Sarah Sloan know your name. Judy smiled back. "It must be nice having a private jet. Not that I know, I've never flown anywhere. I just go to the airport to get patted down. It's cheaper than dating."

Sarah laughed. Lynn cleared her throat, hoping Judy would get the message and return her attention to her, but the reporter in her was paying close attention. Maybe a big story could save her job. And Sarah's husband was running for Governor. And being accused of sexual harassment. So Lynn sat quietly instead of kicking Judy in the shin as she might have done on any other day.

Sarah motioned for Nan to keep teasing her hair. Judy's focus was now squarely on her prey. "So, Sarah," she said, just her first name, as if they were old chums, "Are you ready to be First Lady of Texas?"

Sarah looked perplexed. "Well, it wouldn't be the worst job in the world," she shrugged. "It wouldn't if I didn't have to get the job by putting up with that bastard husband of mine!"

Well, that answers that question, Judy thought.

Well, that's not news, Lynn thought. Everyone in the media knew that Sarah Sloan had been unhappy with her husband ever since the allegations of sexual harassment had surfaced. She still went with him on the campaign trail, but she usually found a reason to leave early and often looked distanced and less-than-happy, even with her "smile for the photographers" face on.

Time stands still for no woman and the clock on the wall kept ticking off the minutes. Lynn realized she was going to be late if Judy didn't finish her bangs soon.

"Judy, I've got to be at the office by one. Sharp!"

Judy reluctantly turned her attention back to her best friend and took her first snip. Lynn's stiff body posture softened a little. There's something about the feel of someone else's hands in your hair that can't help but release some of the tensions of the day. Judy had a theory that people stored stress in their hair. She kept waiting to see it reported in *Science* magazine or *USA Today.*

Nan finished up Sarah's hair, handed her a mirror, and turned the chair around so her client could get a look at the back of her head. The chair squealed. The sound made the hairs on Lynn's arms stand on end. High pitched noises always raised her hackles. Perhaps she'd been a dog in a previous life.

"It'll do," she said. Sarah caught Judy's eyes in the mirror's reflection. "I hate these out of town fundraisers," she said, "But Dick needs the money. I had to cut back on his allowance. He's been a bad boy." Sarah's face, although smiling, was sad. Or exhausted. Maybe both.

Dick Sloan was a squat, ugly little former banker who looked more like an Oompa Loompa than the powerhouse of man he was. His becoming the front-running Republican nominee for Governor was proof of the old adage, "It's who you know." Although in Dick's case it was, "It's what you know about who you know." That's one of the benefits of the banking industry – you stick around long enough, you'll know plenty.

Another client, a slightly heavy-set woman wearing a fake fur coat that looked it, walked into the busy salon, waved to Judy as she passed by, and stopped at the front desk. Donna looked up from her book and yelled toward the back of the salon, "Leroy, your 12:45 is here!"

Leroy emerged from the back room, where he had seated his first client under a hairdryer and had been catching up on the plot on *All My Children*. He strode quickly towards the front desk. As quickly as his too-tight blue jeans, too-narrow ostrich boots, and hand-tooled leather belt with a three-inch longhorn head, horns and all, would let him.

"Hi, sugar. Let's make you beautiful," he whispered as he took his client's arm at the elbow. "Mmmh, what a yummy coat," he said facetiously, although the client took him at his word and smiled.

He walked her back to the shampoo room. "Leroy," she asked "how do you keep from gettin' gored by that big ol' belt buckle?"

"Just one of my many talents. If you're nice, I'll show you some of the others." Leroy removed her coat and led her to a shampoo chair. As he leaned over to suds her he could see up her nostrils. It didn't matter how rich or good looking you were, no one looked good from that angle.

Back at the front desk, the phone rang. Donna picked it up before the second ring. "Lone Star Cuts, home of Texas' biggest hair. What can I do for you?" She paused. "Judy's booked Thursday. How 'bout Nan?" From Donna's giggle, it was clear that the caller thought this was a joke rather than a legitimate offer. "Okay, well, Judy's next opening is, uh, Monday at 9 a.m. Okay, I got you on the book. But, you better bring an extra espresso. Judy's not really awake that early in the morning."

Judy made a mock threatening gesture to Donna. "I'll have you know, I've been getting up at 7:30 and walking with Lynn here every weekend!"

"Is that right?" Donna asked suspiciously.

Lynn, fairly certain at this point that she would be stuck in Lone Star Cuts for the rest of her natural life, nodded. "I can't vouch for her actually being awake, though."

"Hey, I carry on conversations, don't I?" Judy asked defensively.

"You do that in your sleep." Lynn responded and both she and Donna laughed. "Now, cutting hair, on the other hand…"

"I'm working on it." Judy smacked her gum again. "These things take time. You've got to look at it from all angles, take into consideration the shape of your face, the slant of your eyes, the length of your neck. It's an art and I am an artist." Judy almost sounded serious. Until she burst out laughing. Lynn, Donna, and Sarah joined her. Sarah's laugh sounded rusty, like it hadn't been used much lately.

Realizing she had a fairly sizable audience, Judy launched into a comedy bit. "Did I tell you what Nolan did when we were at the grocery store last night?" Lynn shook her head in exasperation. *Maybe I should leave now. I'm sure no one will notice that half my bangs are three-quarters of an inch shorter than the other half,* she thought almost seriously. *I could just grab the scissors away from Judy and finish job myself.*

But Judy was using the scissors to gesture with. "I'm in line and he comes up with a maxi pad stuck over his right eye. 'Look, Mom, I'm Captain Jack Sparrow!' So I kicked him with my peg leg."

Donna laughed. So did Sarah, who was still waiting for Nan to remove her smock.

Lynn shivered. "Boys! I don't know how you do it!" On the list of things Lynn could never imagine herself doing, right after standing in the unemployment line, was raising children. Especially boys. She didn't understand boys at all.

"Well maybe that's why you sleep with women," Judy whispered, maybe a little too loudly.

"Jeez, Judy," Lynn hissed. "Not everyone here knows. Although since you know, I guess I should just assume the worst." It wasn't that Judy couldn't keep a secret. It was just, well, it was that Judy couldn't keep a secret.

Realizing she had trespassed onto sacred ground, Judy switched back to the original question. "It wouldn't be so bad if The Evil One wasn't always late with his child support. I should have a pillow embroidered and sell it on eBay. 'If he's late to the wedding, he'll be late with the child support.' I could make a fortune."

Sarah, finally released from Nan's chair, got up gracefully and glanced in the mirror one last time. Satisfied, she smiled wanly. "Thanks, Nan." She reached into her Gucci handbag and handed a $10 bill to Nan.

"There, done!" Judy exclaimed triumphantly as if trying to impress Sarah with her speed.

"Are you sure?" Lynn asked suspiciously.

A horn sounded. Everyone's attention was drawn to the parking lot, trying to spot the honker.

Sarah knew that horn. It was the horn from hell. The horn that said "Stop messing around and get your ass out here. Because I am the man and you are my bitch." It was a lot for a horn to say, but she heard every word. Sarah glared through the window at her husband's car. He had come to pick her up and take her to the airport personally because she had threatened earlier in the morning to be a no-show.

"Your total comes to $80 today, Mrs. Sloan," Donna said, bringing Sarah's attention back to the business at hand.

"It's a deal." Sarah said as she reached inside her purse for her credit card. "Whaddya figure, hon, $20 an inch?" she joked to Nan, who was now standing behind Donna, reading the appointment calendar over her shoulder.

"What?" Nan asked as she looked up. "Oh. Yeah, we really got it up this time."

"Well, darlin'," Sarah drawled forlornly before shooting a look that could have frozen over hell, or at least Houston, towards the parking lot, "That's the only thing getting up in my life." She displayed a limp little finger to drive her point home.

Donna did her best to display no reaction whatsoever. Judy made a mental note, but also restrained herself. Lynn's eyes widened and she remained in Judy's chair as if superglued in place.

The horn honked again, this time longer, more insistently, as if someone had died and the weight of his lifeless head was pressing on the steering wheel right where the horn was. *If only,* Sarah thought as she threw another look of annoyance towards the door. "Asshole!" she muttered.

Lynn thought, *If I weren't a real journalist, for the next couple of hours at least, I'd be on the phone with my editor calling in a story. 'Dick Sloan: Is his campaign as impotent as he is?' If I didn't have integrity, I'd be raking in the big bucks at a tabloid magazine somewhere.*

Sarah forcefully strode over to the door, opened it halfway and yelled above the chimes, "I'll be there when I get there!" She closed the door and walked back to the front desk slowly and deliberately. "Just think, if he's elected Governor, y'all will get to share this bastard with me!" Although she was looking Donna straight in the eye, the comment was meant for anyone who was listening.

Donna smiled and handed over the receipt. Sarah took it, slid it into her handbag and again slowly, very slowly, inched toward the door, looking out the window to make sure that her husband was aware that she was not at all happy with him. Finally ready to face her foe, she exited.

Through the closed door, the women could hear Sarah's voice from the parking lot. She was standing next to a white Lexus and arguing with the man inside: the infamous, and apparently impotent, Dick Sloan.

Lynn checked the clock again. 12:52. "Oh my God, it's eight minutes 'til. Judy, how can you stretch a five-second trim to a half-hour job?"

"I took a special class in Beauty College," Judy teased. Lynn started unfastening the smock and whisking hairs off the bridge of her own nose. Judy's attention was riveted outside. Finally, she turned to Lynn, who was by now reaching for her purse which had been stashed under the counter. "Besides, wasn't this better than anything that might have happened at the office?"

Lynn nodded. "Sure, it was fun. But I have a job to do. Maybe."

Judy hugged her dour friend. "We're still on for Sunday morning, right?" she asked.

"Yes. Gotta keep these thighs in shape." Lynn had thighs of steel. "God knows, with the way things are at work, I might need to get work as a stripper down at LIVE NUDE GIRLS."

"Always the optimist!" Judy said as she followed Lynn to Donna's desk to check the appointment book. She momentarily stood behind Donna, stretching out her neck. Working gave Judy a pain in the neck. "You sure you don't want to come over later and watch *Finding Nemo* with me and the boys?"

"Judy, as much as I enjoyed that movie the first eight times, no. I'm too stressed out to be good company anyway. Call Mark. I'm sure he'd love to come over."

Judy didn't answer. *Uh-uh. How will Mark know I'm breaking up with him if I keep inviting him to come over?* she thought.

"Now I really have to blow out of here," Lynn said as she paid Donna and handed Judy two ones for a tip.

"Wow, you big tipper. Careful or I might just follow you home."

Lynn was out the door already. Dick's voice was clear as a bell through the opening. "Just get in the damned car and let's go!"

It sounded just like the same argument Judy heard from her neighbors' trailer almost every night. Money can buy a lot of things, but marital bliss isn't one of them. It was clear that there was big trouble in the Sloan household these days. Judy wondered when it would be big enough news to make the papers. But she knew it wouldn't be Lynn who broke the story. Lynn was blessed with fashion sense and high ethical standards.

The front door closed and the phone rang again. Donna had to swivel around in her chair to answer and got the cord wrapped around her neck. "Lone Star Cuts," she said perkily as she attempted to unravel herself.

"Yeah, Ima is working this week. I dunno, I'll ask." She put her hand over the receiver. Her nails were bitten to the quick. Ima was always begging Donna to stop, but it was the only bad habit she had left, and she wasn't giving it up. "Hey, Ima, those Dallas Cowboys nail decals come in yet?"

Ima's childlike voice could barely be heard from the back of the salon, "No. But they'll be here Tuesday."

"Did they call and tell you they'd be here or is that a premonition?"

"Premonition," Ima replied.

Donna removed her hand from the receiver. "Nope. They're not in yet. Try back Tuesday. They should definitely be here by then." Ima stocked all kinds of odd fingernail decals and decorations – football-related, holiday-inspired, even tiny little replicas of famous paintings for the culturally-aware. Donna thought women were crazy to pay up to $200 for these things. But then Donna didn't have their kind of money. Who knows what being that rich did to your mind? Just look at Sarah Sloan, out there arguing with her husband in a parking lot in broad daylight. She looked like a nice normal woman, but who really knew?

Judy had just finished sweeping up as a man walked in the front door. She looked up and smiled.

"So, you Joe?" she asked.

He just pointed to his belt, which spelled out JOE in black letters across the back.

"Well, that's all the ID I need." Judy found it pretty funny that so many Texas men felt the need to identify themselves by their belts. It made her think of the time her mom had used a waterproof marker to mark all her underwear the year she had gone to summer camp. Judy had hated it and was happy they couldn't afford for her to go again.

"I bet that thing comes in handy when you've been out with the boys drinking and you forget who you are," Judy joked as she seated Joe and handed him a smock. Joe gave her a blank stare. Judy looked over at Donna and they simultaneously rolled their eyes. "So, what can I do you for today, Joe?"

Outside in the parking lot, the white Lexus was gone. But Sarah was still there, her face the same color as her hair. She had her cell phone out and was punching in the number of her personal assistant, Trisha. Sarah was seriously contemplating having Trisha take her to her directly to her lawyer's office. *So what if it hurts Dick's career? Who the hell does Dick think he was, anyway? If I want to invest my money in a feminist movie production company, I damned well will,* she thought as she listened to the ring on the other end. She took a deep breath to try to keep from boiling over.

"Just a trim," Joe was saying, although Judy was clearly distracted. "Be careful on top."

Judy stopped staring out the window and looked at the top of Joe's head. Crop hadn't come in this year. Or the past few.

"It's my damn job," Joe continued. "My hair's falling out, my blood pressure's up, some days I think I'm having a heart attack."

"What do you do? You a rodeo clown or something?" Judy asked.

Donna winked at Judy. Sarah Sloan came back inside and sat on the couch near Nan. Although the blustery wind hadn't moved her hair an inch, Judy thought she looked ten years older than when she'd walked outside.

"Anything we can do to help?" Donna offered.

"No thanks, my assistant is on her way," Sarah said, still breathing deeply. *The last thing I need is an ulcer,* she thought. *One oozing hole in my life at a time is enough.*

Oblivious to the scene playing out behind him, Joe shifted in the chair and asked somewhat defensively, "Rodeo clown? What makes you say that?"

"Nothing. I was kidding." Once again, it was apparent to Judy that at least when it came to her sense of humor, men were definitely from Mars, or at least Louisiana. Judy looked over at Sarah again. Her eyes were closed and she was taking deep breaths. It was Sarah's attempt at meditation, but her thoughts were anything but peaceful. Unless envisioning your husband trussed up and roasting over a barbecue could be considered peaceful.

Judy turned her attention back to Joe. He was in his early thirties, perfect teeth and nice hair, what was left of it. Not bad looking. "So what do you really do?"

"I'm a dentist." He said, almost apologetically. He stared blankly at his reflection in the mirror.

"Dentist, huh? Did you ever work on anyone famous?"

He brightened. "Yeah. I did a crown on George Bush. Back left molar, I think."

"Old George or young George?"

"Old. I almost got to do Willie Nelson once, too, but he canceled at the last minute. I think he had an audit"

"That makes sense. I wouldn't want to have dental work and meet with the IRS on the same day." Judy began cutting his hair, careful to leave undisturbed the lonely soldiers at the top of his head.

"I've had my own brushes with fame right here in this chair. I cut former Governor Ann Richards' hair once. Now she had some big hair. I saw a picture of her riding a motorcycle and I thought, with all the spray I put on that hair, she sure wouldn't need a helmet." She paused wistfully. "I miss her."

Judy ran the comb over the left side of Joe's head again. "This is some cowlick you got here. I could mousse it down to hold it in place so I can cut it even with the other side if you want."

"No mousse!" Joe replied forcefully.

"Okay. How about a little duct tape?" Judy asked.

Donna chuckled.

Bernadette dropped by the front desk to look at the appointment book. Her rounded shoulders screamed "Is there a massage therapist nearby?" Seventeen years on your feet cutting hair can do that to a person. "So they're not beating a path to my door today, huh?"

"Guess not," Donna said..

"Okay, then I'll take a smoking break." Bernadette replied as she slapped her forehead in mock forgetfulness. "Oh, I forgot, I gave it up. That Judy is a bad influence." She shot a look at Judy and smiled. "Shit. I guess I'll just go stand outside and fantasize about smoking."

From her perch on the loveseat, Sarah brightened. "I'll go out with you. You can breathe in my smoke while I wait."

"Fabulous!" They exited together, Bernadette making an obscene gesture towards the door chimes as she followed Sarah outside.

Judy wished she was the one sharing Sarah's smoking break. Not that she liked being around smokers, but Sarah was more interesting than the guy in her chair. "So, Joe," she said, bravely pursuing the conversation, "I want you to know that even though I really hate dentists, I am not going to take out my hostilities on your rapidly-thinning hair."

Joe warmed up a little and actually cracked a smile. Yep, the teeth were gorgeous. "I appreciate that." He said. He looked up at Judy, pausing, as if working up his nerve. "Can I ask you something?"

“Sure.”

“Those hair creams that are supposed to make your hair grow back? Any of them work?”

Judy shook her head. “Sorry Joe. It’s all really a bunch of bull hockey. Actually bull hockey would probably grow hair faster than the stuff they sell on infomercials. You could get surgery, but you know what?” Judy paused, then whispered into his ear, “A lot of girls like bald guys.”

“Sure they do,” Joe knew this was just a line nice women used to make balding guys feel better.

“It’s true. Donna there at the front desk. She’s one of them. And me personally, I don’t care what’s on a guy’s head. I care what’s in his heart.” *God, I sound like a Hallmark card*, Judy thought. *Where did that come from? I’m not a Hallmark kind of girl, I’m a Harley kind of girl.*

“Well thanks. I appreciate that,” Joe said, smiling again.

“How about we shave it all off and see what you look like?” Judy offered.

Joe shook his head emphatically.

"Okay, but if you change your mind, you come see me. I think you’d look like Garth.” Joe seemed to think about it for a second, but shook his head a second time.

A commotion arose from Nan’s station. From the corner of her eye, Judy could see Nan’s cart tipping over and as it fell, rollers, combs, and assorted hairaphernalia spilled everywhere. "Now look what you’ve done!" Nan yelled at her reflection in the mirror. "You’re an incompetent boob!” Bawling, she ran to the back of the shop, past six styling stations, the shampoo area, and into the bathroom.

The woman in Nan's chair, a twenty-something student with a barbed wire tattoo around her ankle, sat dazed and confused.

"Not to worry," Judy said, consoling Nan's client, who was looking somewhat freakish with one side of her hair sticking out from a highlighting cap. "Nan's only dangerous when she forgets her Lithium AND her Midol."

The girl began to giggle nervously. Judy whispered to Joe, "Don't worry, I'll keep you safe. I don't like to lose new clients just because my station is next to the Twilight Zone."

Seconds later, Nan returned, having meditated or medicated, or both. She took one look at the mess, stepped agilely over it, and continued pulling her client's hair through the cap as if nothing had happened. "You'll like these highlights," she said cheerily. "They'll bring out the flecks of gold in your natural..."

The rest of her sentence was drowned out by the sound of car tires screeching in the parking lot, followed by a loud horn blast. Judy looked up from Joe's head as Sarah rushed back in the door.

"Everything's okay. I called 911." She stood in the doorway breathless and shaking.

Through the open door a man's voice could be easily heard. "Tell that fucking kid to stay outta my way!"

Sarah caught Judy's eyes. "Judy, I'm sorry…"

Judy rushed out the door. When she got outside, she immediately spotted Nolan, sobbing on the sidewalk. Bernadette was examining him for cuts or bruises. Through the corner of her eye, Judy saw a white Lexus racing down the street.

"Are you all right, baby?" Judy said rushing to Nolan's side.

Nolan, who had been sitting in a crumpled heap, pulled himself up as tall as he could get. "Mom! I told you I'm not a baby!"

"He seems to be okay," Bernadette offered. "Sarah called 911 just in case."

Judy stared down the road where the Lexus had disappeared. "I know you're not a baby." She hugged Nolan tightly.

"Mom!" Nolan pulled away and pointed to his bike, the front wheel completely bent. "My bike!"

"We'll take care of your bike. How about you?" *When did he stop letting her hug him*? "Are you all right?"

"Yeah, sure." Nolan stood up and kicked the front tire of his bike.

Judy, still squatting, brushed the dirt from his jeans. "Next time, you come right inside when you get here, okay?" He just shrugged.

"What happened?" Judy asked Bernadette, reassured that Nolan was in one piece.

"I'm not sure. Sarah was smoking her cigarette and we were talking about the bats under the Capitol Street Bridge. Then HE drove up again. She went out to the car and they said something to each other and then she turned around like she was going to go back inside. He gunned the engine and before I knew what had happened, Nolan came flying through the air and landed on the ground where you found him."

"Asshole!" Judy said.

"Asshole," Nolan repeated.

"Nolan!" Judy scolded.

Sarah reemerged from the shop. "Judy, I'm so sorry. I can't believe he…"

Judy looked Sarah square in the eye. "It's not your fault." Judy knew all about asshole husbands. There wasn't anything you could do about them, no matter how hard you tried or, apparently, how much money you had.

Sarah didn't look relieved. "I'm the one he was mad at. And I know he gets like this when he's mad. One of these days I'm gonna…"

"Why don't we just go back inside? It's windy out here," Bernadette suggested, her teeth chattering.

"Come on, Nolan. Leave your bike out here" Judy said as she nudged her son toward the door. This time he let her touch him without pulling back right away. Judy noticed that Sarah hesitated to join them, so she used her other hand to grab her shoulders and push her towards to door as well. "Come inside. You don't want to catch pneumonia. And you don't want the wind to mess up your lovely new do."

"A flying mobile home in a twister couldn't move this hair, hon," Sarah said, attempting to joke away her anger. She was glad Judy wasn't holding her responsible, but her blood pressure was still higher than Exxon's quarterly earnings. Once inside, Sarah apologized again. "I'm sorry, Judy." She looked down at Nolan. "Don't you worry, I'll buy you a brand new bike."

"Cool. Can I get an electric scooter instead? Can I, mom?"

"Honey," Judy interrupted. "It's not nice to be greedy." To Sarah she said, "Just a regular bike like the one he had will be fine."

Nolan groaned.

Sarah opened her purse and pulled out her checkbook. Before Judy could stop her, she wrote a check for $1000. She handed it to Judy.

"Sarah, that's way too much. I only paid $100 for the one that Dick ran over out there."

"But Mom, this way I can get a good bike and a scooter. And I promise I'll always wear my helmet." Nolan knew how to push her buttons all right.

"We'll see," Judy said. Then, to Sarah, "Thanks. You

really didn't have to." Judy was already calculating what she could do with the money that would be left over. Maybe get someone to put weather stripping around the windows of the trailer. Or get a new dishwasher. The one she had now seemed to put more food on the dishes than it took off.

"Yes I did," Sarah said patting Judy on the shoulder. "I threw in a little extra for pain and suffering."

Judy couldn't help but laugh. "Yours or mine?"

Nice Guys Last

There's something about a run-in with a really bad boy, a boy like Dick Sloan, that can make a girl like Judy reconsider her predilection for the dark side of manhood.

After the paramedics checked out Nolan and Judy took him home, she was still shaken from his near-death experience. Maybe it wasn't a true near-death experience – he hadn't, after all, reported seeing a white light or being beckoned by all his rats, hamsters, gerbils, turtles, and frogs that had gone on to animal heaven. But in Judy's mind today's accident was close enough. She loved her boys like crazy. Sometimes literally, dressing up in silly costumes and playing games with them. Not so much lately, because they thought she should grow up. But she kept the pirate's eye patch and the funny glasses with the eyes that dangled in her dresser just in case they ever needed her to be silly again.

Judy had to talk to someone and de-stress, so she called Lynn.

"You've reached Lynn," the answering machine said. "I'm sorry I missed your call. Please leave a message."

Judy had tried to get Lynn to put something more interesting on her answering machine. She'd even written a couple of answering machine messages she thought would spice up her love life, but Lynn preferred straightforward and boring.

Damn, Judy thought, as her fingers started dialing Mark's number without her brain having a chance to think. As it rang, she realized what she'd done. *This will throw a monkey wrench into my whole plan*. She'd been careful not to let him think their relationship had a future. But he hadn't seemed swayed. He liked her. She made him laugh. And she wasn't afraid to be completely, totally herself. That was a refreshing change from the people he spent his days with – politicians and lobbyists and staffers – all of whom spent most of their time trying to pretend they were more important than they really were.

"Hi, Mark." *We can still be friends*, she told herself as she heard his voice on the other end. *It doesn't have to be love. I only called him because my other friend Lynn didn't pick up. And I'm not in love with her either.* Rationalization is a funny thing. The more thought you put into it, the further off base you usually are.

"Hi Judy! Wow, this is nice." He seemed genuinely happy to have her call him instead of vice-versa. "What's up?"

Judy's resolve to keep her emotions in check since Nolan's run-in with Dick shattered into a thousand tears. She sobbed into the phone. "Nolan almost got killed today."

Without hesitation, Mark replied, "I'll be right over."

Twenty-five minutes later, Mark knocked on Judy's door. By then she had wiped the mascara streaks off her face and changed out of her sweat pants and t-shirt into a skirt and blouse. She cleaned up the living room and lit a candle, all the while telling herself she wasn't preparing for a romantic interlude, she was just tidying up as she would if Lynn or Donna or anyone else were coming over.

Judy opened the door. Mark stood there holding a handful of yellow tulips. "It's all they had at Safeway," he said as handed them to Judy. "I didn't want to run around looking for better flowers when you needed to talk."

"They're beautiful," Judy said quietly. Suddenly, the day's emotions washed over her again and she found herself in his arms.

Ten minutes later, after telling him the story of Sarah and Dastardly Dick, and watching Mark getting outraged, but not scary-outraged, they checked on the boys together. Nolan and Stevie Ray were both sound asleep and looked no worse for wear. And as Mark headed back toward the living room, Judy took his arm and gently led him into her bedroom. *Damn*, she thought, this *is really going to screw things up*.

And it did. He held her tenderly and kissed her neck softly. Slowly he unbuttoned her blouse, drinking in every part of her like a man who was dying of thirst. He ran his soft hands over her warm skin, paying attention to her back, her waist, and her shoulders – not zeroing in on her breasts and pussy right away like some kind of heat-seeking missile that couldn't afford to be off course for even a second. *This must be what foreplay is like*, Judy thought, remembering clearly that neither Clay nor Juan had ever spent this kind of time warming her up and turning her on.

And when she was ready, she was good and ready. She said a silent prayer that in this light, he couldn't see her cellulite, then unzipped Mark's pants, wrapped her legs around his hips and took him.

She came three times. And she came to her senses, deciding that a good man wasn't such a bad thing after all.

Help, I'm Turning into My Mother

Almost every morning for the past year or so, when Lynn looked in the mirror, she saw her mother and got depressed. Her mother was wonderful – she'd not only been okay when Lynn came out to the family, but had tried to set her up on dates with the lesbian children of her friends. But no woman likes it when her mother's reflection stares back at her.

Sarah Sloan never saw her mother in the mirror because she'd been adopted and never knew what her real mother's face looked like. Sometimes she thought this was a good thing, not having to worry about the bad habits or medical conditions her mother might have. Other times, especially lately, she wondered what her natural mother was like. But Sarah was extremely thankful for adoptive mother and one thing she had given her that made life much more livable: cojones.

Ima saw her mother in the mirror all the time. But it was her mother from a previous lifetime, and she was an Egyptian queen. That made Ima happy because at least once in her life history, she'd been a princess.

Judy wished she could see her mom's smiling face in the mirror staring back at her. The mom Judy remembered best was a woman who was tired a lot, whose hair fell out from the chemotherapy and then came back in darker and curlier, and fell out again. A woman whose voice sounded raspy from the radiation therapy after the cancer moved from her breasts into the nodes in her neck. Judy wished she could remember the mom in the pictures on her bedside table and in the albums – the laughing, energetic woman, who had loved having four daughters and staying at home raising them. The mom who was beautiful without make-up, something Judy had never felt.

The last day of Judy's mother's life had brought enormous sadness, but Judy also felt relieved. No teenage daughter should have to see her mother suffer that kind of pain, and being the eldest, Judy was the one who had gotten a provisional driver's license to drive her mom to her appointments, who had held her hand when the chemo needles went in, and who had been there when the doctor explained that things weren't going as well as he hoped. And Judy had been the one who had tried to help her younger sisters, Joy, Jas, and Jenny, accept that when they put mom in a box in the ground, she wouldn't be coming back. Joy and Jasmine, at ages 10 and 12, knew what death was. Jenny was only five though, and the hope that lit up her face every time the front door opened broke Judy's heart.

Even though Judy's mom's face didn't stare back at her in the mirror, she was there nonetheless – every time Judy chose her boys' happiness over her own. Every year when the four sisters and Judy's dad would get together for a family reunion and cobble together their mom's memory. Every day Judy made wild fashion choices, despite the fact that she could only remember her mom in a faded cotton hospital gowns that snapped up the back. And her mom was definitely there every time Judy told a funny story. Those laugh lines that were just now forming around Judy's well-mascaraed eyes were definitely her mom's.

I'll have Two Breasts and Some Fries, Hold the Mammogram

A woman with large, firm breasts buttoned tightly into a short sweater the color of Pepto Bismol turned the pages of a three-year old *Family Circle* magazine, not really reading. Her sweater blended nicely with the turquoise and rose pattern of the sofa she sat on, as if she had coordinated her outfit to match the furnishings. Maybe she was trying to disappear into the background so the nurses and technicians would go on about their business and leave her alone. An attempt at mammoflage, perhaps.

She flipped another page and sighed. Only the goldfish swimming worry-free in the tank fifteen feet behind her heard anything. And since no one would be squeezing their breasts flat in a few minutes, they thought nothing of it.

Judy walked up to the door of the West Austin Mammogram Clinic and hesitated. Bad memories overwhelmed her and she almost turned around and left. It takes a lot of strength and courage for any woman who has lost a mother, sister, aunt, or a friend to walk through those doors. Judy bit her lower lip, reminded herself that Stevie Ray and Nolan needed a mother, at least to keep them from killing each other for the next five years and to embarrass them in front of their friends, and pushed through the doors with false bravado.

Once inside, Judy stood for a few seconds, summoning the gumption to check in. She took a look around and noticed the women sitting in the waiting area, including the woman in the pink sweater, then walked to the reception desk to sign in and get the whole thing over with.

"Mrs. Porretto," a nurse called from a doorway leading to a maze of dressing rooms. An older woman, perhaps in her seventies, practically leapt out of her seat on the other side of the fish tank. She followed the nurse down the hall obediently like a naughty second-grader making the long trek to the principal's office once again. Judy was glad the woman in the pink sweater was still there. Someone to talk to.

"Have you been here before?" the woman behind the desk asked in her most professional voice. *It must be hard doing her job*, Judy thought. *All these women coming in, nervous and anxious, some of them already knowing they have cancer, others who will be finding out today. I hope I'm not one of those. Please God, let my mammogram normal.*

"Yes, last year." Judy remembered it like it was yesterday.

"Your name?"

"Judy. Judy Parker."

They both waited for the computer to search for her records. "Here you are. Judy Parker." Judy started tapping her fingernail nervously on the counter top. The receptionist looked up. "Do you still live at 103 Riverview?"

"Yes," Judy replied, noticing just how tiny the receptionist's breasts were. There was something about being in a mammography clinic that made it hard to take your eyes off other women's breasts.

"Do you still have the same health plan?"

"Yep." April had started providing everyone at Lone Start Cuts health insurance two years ago. Until then, every time the boys had scraped a knee or gotten a sniffle, Judy had worried that she'd have to sell a kidney to afford the doctor bill. Fortunately, Stevie Ray and Nolan had been pretty healthy, Judy too. But things change, so every night Judy included April in her prayers – both for being the kind of boss who would put up with her and for making it possible for her to keep her family healthy.

The receptionist looked up and smiled politely. Judy knew that smile might be the only good moment in some woman's visit here today. "You can have a seat now, Judy," the receptionist interrupted. "A nurse will be with you in just a few minutes."

Judy turned and walked towards the woman in the pink sweater and sat down across from her, trying not to stare at her ample chest. She picked up an old *Glamour* magazine and thumbed through it.

"Oooh, goody, the 1997 fashions!" Judy said loud enough for the woman to hear. "Looks like I'm finally in style."

Pink Sweater looked up. She had exchanged the *Family Circle* for one of the clinic's mauve brochures. "It could be worse. You could be reading this," she said as she held up the pamphlet. Judy could just read the title, "Healthy Breasts Start Here."

"Oh, now that looks interesting," Judy said smiling, glad to have found someone willing to chat with her so she could get her mind off things. "Is there a healthy breast diet and exercise program in there? Because I've been wondering whether I should put mine on low carb or low protein."

Pink Sweater laughed. "Nope, nothing as interesting as that." Her eyes met Judy's. "Just the same ol' stuff – self exams, regular breast exams by your doctor, yearly mammograms after 40. The things we all know, but they like to put down on paper to make sure we don't sue them if anything happens."

Please, don't let anything happen, Judy prayed silently. "Anything about whether underwire bras cause cancer? Or deodorant?" Judy had heard about both of those on the radio, but she wasn't going to give up either unless there was damned good evidence.

"Doesn't say." Pink Sweater put down the brochure. She lowered her voice. "I hate coming here. I know it's silly. It doesn't hurt that much. I've never had an abnormal mammogram, but every time I come through that door, my blood pressure shoots up and I get a raging migraine."

Judy nodded sympathetically. "I know what you mean. Who needs horror movies when we have 'The Mammogram?' Freddy Kruger actually performed my last mammogram and I have the scars to prove it."

Pink Sweater laughed. Judy smiled appreciatively. "I tried to get my friend Lynn to skip work and come with me today and hold my hand. But she's a workaholic. And more scared of this place than we are." Judy got up and moved closer to Pink Sweater. "I guess for today, you're my best friend. My name is Judy."

"Hi, best friend Judy. I'm Debbie," she said not at all fazed by Judy's presumptuousness. "But as your best friend, there are a few things I should know. Like how old are you?"

"Thirty-three."

"That's young. I'm forty-two."

"Wow, you look great. And those breasts of yours, they're spectacular."

"Thanks, they should be with what I spent for them." She winked at Judy. "I got them for myself two years ago as a birthday present."

"They are remarkable. I'm sure your husband..." Judy checked Debbie's finger for a ring. "Or your boyfriend is very happy."

"Neither."

"Oh."

"I'm taking a break from men. I was married for fourteen years and that's over, so now it's Me Time."

"Good for you."

Debbie turned her head to the side to get a better look at Judy. "Thirty-three is kind of young for mammograms, isn't it?"

Judy nodded, suddenly remembering again why they were both there. "My mom died of breast cancer. And my doctor found a lump. I tried to tell her that THAT was my breast."

Debbie shook her head sympathetically.

"She's pretty sure it's nothing. I just have lumpy breasts. Which makes sense since the rest of me is lumpy too."

"You don't look lumpy to me," Debbie said matter-of-factly.

"Thanks. I have lost a lot of weight recently. Some of the lumps are scattered about in various parks and parking lots across town."

"Congratulations," Debbie said with the appreciation of someone who knows how difficult losing weight can be.

Another nurse emerged from the long hallway and read out another name. Judy and Debbie braced themselves. "Gina Jimenez?"

A middle-aged woman got up and followed the nurse to the dressing rooms.

"Reprieve," Judy whispered.

"What do you do for a living?" Debbie asked, just as thankful for it not to be her turn yet.

“Hair. But I want to be a stand-up comic some day.” Judy figured the more people she told – friends, strangers, homeless guys on the corner holding signs – the more likely it was she’d work up the nerve to actually do it one day.

“If that’s what you want, you should do it,” Debbie said enthusiastically. “Take it from someone who is older, there’s no time like the present.”

Yet another nurse stepped into the waiting room and called out a name. "Judy Parker."

Judy inhaled deeply. “Hey, that’s not fair, you were here first.” She stood up anyway. “Promise me if I come back with a hickey on my nipple, you’ll still respect me in the morning.”

“You got it. Good luck in there.”

"You too.” Judy followed the nurse through the hallway filled with artwork in shades of turquoise and rose, back to the maze of dressing rooms. Instead of a door, each room had a turquoise and mauve shower curtain (not the plastic kind, but the fancy cloth ones that Judy always wanted but could never see paying $60 for, not when Stevie Ray and Nolan would probably end up using it as a parachute to jump off the roof with.) Judy wondered if there was a store in the mall that only sold things that were turquoise and mauve.

The nurse, whose breasts were average-sized, took a quick look at Judy and pulled out a pink cotton gown, sized M. "Here, this should be about your size."

Despite her anxiety, pride surged through Judy. A complete stranger had looked her over and guessed she was an M! Not an L or an XL, but an M! Of all the wonderful moments Judy had had in her weight loss process, this was right near the very top. An M! What she wouldn’t have given to be an M back in high school.

"The opening goes in the front," the nurse was saying, interrupting Judy's revelry. "Please remove everything above your waist and be sure to take off your earrings." Judy noticed how the nurse's breasts swayed as she talked. Like a metronome keeping time to the cadence of her conversation.

Judy immediately removed her earrings and stashed them in her purse.

"When you're ready, bring your things and meet me right down the hall by the lockers."

It wasn't until after the nurse left and closed the curtain behind her, that Judy noticed the new age music being piped in to the dressing room. She recognized it -- Enya. Lynn had the CD which she played on her expensive Bose stereo system. Judy had an old phonograph she'd bought at a garage sale, along with speakers the size of refrigerator boxes she'd gotten at Goodwill after she threw Clay out. The speakers took up practically the whole living room. When they were little, the boys had piled sheets over them and played fort.

Judy quickly removed her sweater and bra and wrapped herself in the pink gown. She headed towards the locker with her belongings. The nurse showed her which locker to use and handed her the key, which was attached to a pink, spring-like plastic wrist band. Not quite rose, but very, very close.

Together, Judy and the nurse walked toward the waiting room outside the mammography area. Two other women were there watching a video on self examination. Judy turned away. She didn't do monthly self exams. It was too scary. She knew it was stupid, in light of her family history, but Judy had had enough scares as a teenager. That's why she avoided crime shows on TV, scary movies, and looking for lumps or spiders in the shower.

Just as she was about to take a seat, a tall redhead with shoulder length hair flipped under on the ends, and large, but clearly real, breasts called her name.

"Judy Parker?"

"That's me," Judy answered anxiously as she got up and took a few steps toward the redhead.

"Come on in. I've been waiting for you all day," the technician said, cajoling Judy into the radiology room.

As she got closer, Judy read her name tag: "Gail." Gail was also wearing a button on her turquoise scrub that said, "Smile, you're on candid x-ray!" and a silver chain that nestled between her breasts. What Judy couldn't see was that at the end of the chain was a silver heart that said "Dog Mom."

"Hi Judy, I'm Gail. I'll be your server today."

As she stepped into the room, Judy nodded nervously, suddenly at a loss for words.

"We'll try to make this as painless as possible," Gail continued.

"We?" Judy asked, looking around to see if there was anyone else in the room. Just the two of them.

"Me and whoever it is playing this damned peaceful music," Gail offered, as she looked at Judy's chart. "I see you've done this before. So you know the drill, but just in case you've forgotten, I'm here to remind you."

Judy nodded again mutely. *A mime has a mammogram*, she thought as she tried to summon more words to come.

Gail led Judy over to the machine. "Put your feet on the footprints and we'll get started. I'll need you to unfasten your gown and drop the left side."

"What, no dinner and a movie first?" Judy blurted out, more from nerves than from the desire to be funny.

Gail smiled. Judy noticed that when she did, all her teeth showed. Like Whitney Houston or Julia Roberts. *Probably a good laugher*, Judy thought.

"Well, if you're good, I'll give you some chocolate when we've finished this up," Gail said.

"Okay. But you should know I don't like the gooey pink centers."

"Me either, what is with those anyway?" Gail grinned toothily. "Now, what I have to do is place your breast between these two plates..." She broke off mid-sentence as she gently lifted Judy's breast onto the bottom plate and pressed a button on the machine.

"Your fingers feel like icicles," Judy noted.

"Thanks for noticing. I keep them in the freezer between patients," Gail said laughing. It was a beautiful laugh.

Judy immediately felt better. Well, at least every part of her that wasn't about to be folded, stapled, and mutilated felt better.

Gail began pulling the skin from Judy's back towards her front so that every ounce of breast tissue was on the plate. Judy grimaced. "Sorry. I know this is uncomfortable. Believe me, I do." Gail lowered the top plate of the machine.

Judy looked away from her own breast which was about to pancake between the clear plastic plates. She noticed a coffee mug on the counter that said "Why let reality ruin your day?" *Appropriate*, she thought.

Gail got her attention again by placing an icy finger on a different warm part of Judy's body. "Now here's the best part. I need to pull your breast as far away from your ribs as humanly possible."

"I tell you what," Judy said, getting into the swing of the conversation, "I'll leave my breasts here and go wait in the lobby. Call me when you're done."

"No, you should stay," Gail said, still making adjustments. "You're funny. So many of my patients are sticks in the mud when they're in here. I have no idea why."

They both laughed.

"As long as we understand our roles here. I make you laugh and you cause me excruciating pain. Do I have that right?"

Gail nodded, then lowered the plates further. Had Judy been looking, she would have seen something that looked a lot like road kill, but Judy's eyes were now closed. She was trying to visualize herself somewhere else. Home, work, a desert island with George Clooney. But for some reason he was squeezing her breast flat in his hands. The discomfort jolted Judy back to reality, despite the sentiment on the coffee cup.

Gail walked around behind the glass wall and told Judy to stop breathing. The machine made a whirring sound and the plates automatically released. Judy didn't have to be told to breathe as they did so, it was automatic. She wondered if part of her lung had been caught in there.

Gail returned to Judy's side and pressed a button that changed the angle of the plates.

"You should use this device for good, not evil," Judy offered, breathing deeply.

"You got a suggestion?"

"Yeah, Testegrams!" Judy said, thinking it was a clever idea.

"You wouldn't believe how many women suggest that when they're here. Equality of the sexes, I guess," Gail said as she grabbed Judy's breast again and started laying it on the plate.

"Damn, and here I thought I'd come up with a great idea," Judy lamented, disappointed in someone else beating her to a punch line.

Gail made one final adjustment to the machine. "I'm available to do Testegrams, but so far, I've had no volunteers. Okay, let's take this picture." She walked behind the glass wall. "Now hold your breath." The machine went through its machinations again.

"You can breathe now," said as she came back into the room. As the plates released Judy's breast this time, she looked. Amazingly, her breast popped right back into shape.

"Maybe you shouldn't wait for volunteers," Judy said as Gail began readjusting the machine for Judy's other breast. "I could start a list of guys who deserve to be guinea pigs. First, my ex. Then this guy who keeps harassing my friend Lyn at work. Oh, and of course, let's not forget that scumbag, Dick Sloan."

"Dick Sloan? The guy running for Governor?" Gail's warm voice cooled rapidly at the mention of his name.

"Yeah, him. He practically ran over my son last week. I guess he figured Nolan was too young to vote."

"That's awful!" Gail looked genuinely concerned as she turned Judy's body slightly towards the wall. "Is he okay? Your son, I mean?"

"Yes, no thanks to Mr. Sloan, who sped off in his car," Judy spat. "Fortunately my kids tend to bounce back from being hit by a car like it's nothing."

"I'm glad he's okay. What a jerk. Hard to believe he's running for Governor!" Gail seemed as angry as if Dick had tried to run over her own children. "Sometimes I wonder if they prescreen people for asshole-itis before they let them run for office in this state."

"Asshole-itis? I love that, can I use it?" Judy asked.

"Sure," Gail answered looking a lot less cheerful than she had just seconds earlier. "Okay, you know the drill. Let's get this one done."

Judy obligingly held her tongue and her breath as Gail took another x-ray.

Gail changed the angle once again. Looking away this time Judy noticed a picture on the other side of the room. "Is that your husband?"

"Yep. That's Glen. And our dog kids. Or as my dad calls them, Rodents of Unusual Size. We tried to have kids of the non-canine variety, but it never happened. Now I'm over forty… Anyway. Arm up." She reached behind Judy to grab any extra skin to get a better shot. "At least my dogs won't ever come home with their navels pierced or their hair dyed to look like the state flag."

"I can do that. The hair, not the piercing," Judy said. "If your dogs ever change their minds."

A smile broke out on Gail's face again. Her smile was like the sunrise, warming everything around it.

"Okay, last shot. Hold your breath." Gail disappeared once again behind the partition. "Okay, you can breathe." She returned seconds later. "You do hair?"

"Yep, I do hair. And what do you tell people when they ask what you do for a living?"

"I say I feel women up," Gail said with a giggle. "Okay, you're done. As much fun as this has been, you're free to go."

"Wait a minute. You get a girl naked and then you shove her out the door? You probably won't even call." Not that Judy wouldn't be happy to leave, but she liked Gail. "And what about my chocolate?"

Gail went over to her mug and pulled out a Hershey's kiss. "Voila!" she said, as she presented the gift to Judy, who unwrapped it and popped it into her mouth. No sense carrying it back to the locker. Besides, after what she'd just been through, she deserved chocolate.

"I'm not allowed to call. At least not about your results. The radiologist will call your doctor in three to five days," Gail said. "Okay now, I've got other people who can't wait to have me squeeze their breasts."

Judy started out the door, but she turned at the last moment and looked at Gail. "You should drop by the salon sometime. I work at Lone Star Cuts."

"I may just do that," Gail said, running her fingers through her hair.

Judy opened the door. "I'll do your hair. And we'll dream up ways to fix what's wrong with the world. I've got scissors and you've got all the required parts for the Testegram."

Later that night when Judy got back to her trailer, she found a potted Gerbera daisy sitting on the welcome mat. The note attached to the pot said, "I hope your mammogram went okay. Try not to worry." *That Lynn, she remembered*, Judy thought. But the card was signed "Love, Mark."

It's Not Who You Know, It's Whose E-Mail You Know

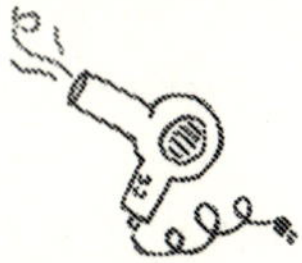

AOL was taking too long to load. *I really should get DSL or cable, because AOL without DSL is aggravating my PMS,* Lynn thought. *Especially now that I have a real reason to be online.* She grinned through her impatience.

That reason had a name, and it was *Snowbunny.* The name was way too cute, the kind that usually made Lynn snort and wonder what got into people thinking up cutesy monikers for themselves. Lynn's user name on Gaydate.com was Reporter13. But despite the *Snowbunny* handle, whoever she was seemed perfect – she liked outdoor activities, exercise, old movies, and reading the newspaper on Sunday mornings. It was the latter that first caught Lynn's attention.

But it was *Snowbunny's* picture that really changed Lynn's mind, not to mention her heart. She was a cute freckle-faced kid, although her profile said she was thirty-six. Her eyes and her smile were proof enough that she had a kind heart. Then there were the two black labs she had her arms around in the picture. *Anyone who can love Labrador Retrievers might be able to love me*, Lynn thought.

They'd been e-mailing for three weeks now. Lynn hadn't told anyone, not even Judy, although there had been plenty of opportunity during their weekend walks. Judy would be ticked off when she found out Lynn had been hiding this secret from her. Judy didn't believe in secrets, but that was mostly because Judy couldn't keep her mouth shut, so she didn't think anyone else should either. But Lynn didn't want to jinx things. *Snowbunny* could still turn out to be a sixty-year old retired truck driver who liked to yank lesbians around instead of doing porn. She hoped not. And even though Lynn was usually pessimistic about, well, everything, she believed in her heart that Snowbunny was exactly who she seemed to be. And that made Lynn happy.

"You've got mail," the computer voice announced. Lynn jolted to attention. With a tiny flutter of her not-so-hardened heart, she saw that noticed Snowbunny was online.

"Hi, Snowbunny. How was your day," she typed as quickly as her fingers would go. At least being a reporter gave her that advantage. She could type faster than the rest of the women who might be trying to hit on the cute little online rabbit.

SOBs and Other Sob Stories

As Sarah Sloan drove up to her husband's campaign headquarters, she prayed she wouldn't have to go inside. She was in no mood to put up with him today. Her heartburn was already setting fire to her esophagus, and the doctor told her she really didn't need any more stress in her life. As if there were people who did. The pain in her stomach and throat felt like it might burn through her skin and set her jade green jacket on fire, which would be a damned shame because that jacket matched her eyes like they were from the same dye lot.

In Sarah's mail box that morning, wedged between the shop-by-mail catalogs and the invitations to Austin social and political soirees, of which she was rapidly tiring, was this month's bank statement. Having a bit of an anal-retentive streak, she dropped the rest of the mail on the pine buffet and tore open the bank envelope. Her eyes scanned the statement, stopping abruptly when they reached what they were looking for: Dick had made seventeen withdrawals totaling over $300,000 written out to the Dick Sloan for Governor Committee. Her hackles stood up. On a clear day, you could see those bright orange hackles all the way to San Marcos.

"Why don't you just divorce the son of a bitch?" was a question Sarah heard almost daily from her closest friends. They didn't understand why she stayed with a man who dwarfed her intellectually, socially, financially, and in stature. Had he had been one of the Seven Dwarfs, he would have been Grumpy's stand-in, Malevolent.

Sarah, on the other hand, had graduated first in her class from The University of Texas School of Business, and had quickly taken a little seed money from her parents and built *She Said*, the biggest PR firm in the south.

"You shouldn't waste any more of your life on the bastard than you already have," her friends insisted. She agreed with them, but something inexplicable kept her by his side. For a long time, she stayed with Dick because of Jen. But now their daughter was practically grown. Some days, she stayed because she thought it was just too much trouble to start over. At least Dick was the devil she knew.

When she looked at it analytically, with the pros and cons listed on a spreadsheet, Sarah could see clearly that staying in a marriage with an egotistical maniac with a vicious mean streak and a total lack of empathy had no pros. Therefore she must have some kind of a genetic defect. Her mother, who these days just looked at her and shook her head sadly, had stayed married to Sarah's dad for forty-two years, until he died of cirrhosis of the liver. And grandma, who'd been married to grandpa for 71 years, spent sixty of them not only in separate beds, but in separate states.

Maybe what kept Sarah married to Dick was some kind of emotional superglue that had been applied back in high school before she was old enough to know better. Even now, she would still have occasional flashes of the Dick she knew in the eleventh grade. Moody and dangerous, but fun. Dick was still moody and dangerous, but Sarah had grown up. Women often do.

Sarah pulled into the parking lot, turned off the ignition and looked around. She had hoped Jen would be standing outside waiting for her when she drove up, but no such luck. She'd have to go inside. *Damn it!* she thought, praying that Dick was out sucking up to cattlemen or kissing babies or whatever he spent his lunch hour doing. Not that she had any desire to know. Too much information, in this case, equaled too much heartburn.

She glanced through the glass door, but saw neither her daughter nor her spouse, so she snuck inside. The large room was muggy, muggier than outside. She looked around. No Jennifer. Just campaign workers busy answering phones and stapling yard signs onto stakes. Sarah couldn't understand why her daughter, only nineteen, had chosen to spend spring break playing indentured servant to her dad. *Just further proof that blood is thicker than water*, she thought. *And messier.* She snickered.

Today was their monthly mother-daughter lunch thing, as Jen called it. Although in Sarah's mind they were more like Thelma and Louise than mother and daughter. Only neither of them had worked up the nerve to leave town yet, but Sarah was definitely up for a joyride. Even if it meant killing a man and pitching him head first off a cliff. She had the man all picked out.

"Hey, Sarah, what‘chya doin' here?”

Sarah would know that gravelly voice anywhere – Jimmy Joe, Dick's campaign manager. His twenty plus-year tenure in smoke-filled rooms of the hallowed halls of the Texas Legislature before they had gone smoke-free had left his voice as raw as the slab of beef he regularly consumed at dinner.

She turned around and saw Jimmy Joe clutching a fat report with his oversized fingers. The man had hands like a grizzly. Sarah wouldn't have been surprised to see him clutching a freshly-caught salmon instead of the results of yesterday's polls.

"Hey, there, Jimmy Joe. I'm here to pick up Jen. Seen her anywhere?"

Jimmy Joe looked tired, more so than usual. Which was probably to be expected this close to the primaries. But it was more than exhaustion registering on his haggard face, which matched perfectly the pale puke color of his wrinkled dress shirt. Loss of spirit was more like it. "She ran out to The County Line to git the volunteers Dr. Peppers and barbecue. Should be back real soon," he answered slowly.

"Thanks, Jimmy Joe." Sarah smiled comfortingly. "And you hang in there."

"Thanks, Sarah. You too." Jimmy Joe looked at the report for a second, then looked up at Sarah. "He needs you, Sarah. He may not act like it, but he does."

"He needs my money, Jimmy Joe. There's a difference." *Dick doesn't deserve Jimmy Joe*, Sarah thought. Hell, he didn't deserve either one of them. But somehow, he still had them both.

The phone on the desk next to Sarah rang. One of the volunteers stopped stuffing bumperstickers into manila envelopes and picked it up by the third ring. "Sloan headquarters. A Big Man for a Big Job. How may I help you?"

Jimmy Joe wandered off, lost in the numbers. Sarah glanced furtively towards the window, hoping Jen would show up before her father.

"If your property is on a corner, we suggest two yard signs. That way nobody misses them. Yeah, pick 'em up here. 1812 Nueces. Someone will be here..." He put his hand over the mouthpiece and yelled across the room, "Hey, Vic, when are you leaving?"

Another volunteer, dressed in an Arnold Palmer golf shirt and green checkered double knit pants, looked up from stapling signs. "Around 10. If I make it that long."

"Get here by ten," the volunteer suggested before hanging up the phone.

"Hey, Mom! You're early!" Jen's voice rose above the buzz of the room.

Sarah turned to see her daughter wrestling with the door, her hands full with a tray of soft drinks and three bags of food. Like her mother, Jennifer was impeccably dressed, coifed and manicured. There was, however, the tiniest of barbecue stains on the cuff of her left sleeve. No one but Sarah noticed.

"I thought we were going to lunch," Sarah said, pointing towards the County Line bags and acting hurt.

"We are, Mom," Jennifer sighed with a touch of mock exaggeration. "I was just getting food for the guys. It's what I do around here."

"So that's what you do," Sarah said trying to keep the sarcasm out of her voice. That was for Dick, not her daughter. "I thought you'd be in charge of the whole shebang by now. Lord knows you're smarter than the guy in charge."

"Hush!" Jen said forcefully. "Just because you and dad can't be in the same room with each other doesn't mean I have to avoid him. He needs all the help he can get."

Sarah smiled and winked at her daughter. "You're right about that. But not the kind of help you or I can give him. A five-year stint on a therapist's sofa maybe."

"Mom! You promised!" Jen unloaded her packages on a desk half-covered with direct mail flyers waiting to be hauled down to the post office. Another of her chores.

"Okay, okay," Sarah said. "Let's get out of here."

But before they could make their escape, Dick came through the front door. He didn't notice Sarah. A momentary blessing for them both.

"Jimmy Joe, where the hell are those Dallas numbers? Goddamn it! Why do I have all these morons working here," he gesticulated wildly, "when I have to do everything myself?"

Jimmy Joe reappeared from his office and handed Dick the report he'd been scouring. "They're right here, Dick. Keep your shorts on."

"I see you're in a good mood for a change," Sarah chided her husband.

Dick turned to face his wife with the joy of a man turning to face his firing squad. "I didn't know you were here. What do you want?" he asked rudely.

"Jen and I are having lunch," Sarah said, as if to clarify that she was in no way here on his behalf.

Dick slammed his fist down. It landed right on top of one of the barbecue sandwiches. Sauce went flying everywhere. Sarah and Jen both jumped instinctively out of the way. Dick didn't seem to notice the sauce or their reactions. "Goddamn it Sarah, you know…"

Ever the peacemaker, Jimmy Joe interrupted. "Dick, it'll be all right. Things are going smoothly. We can let Jen have a few hours for lunch."

"Yeah, Dad. Unless there's a Dr. Pepper emergency in the next hour, I think the office will hold up." Jen grabbed her Endangered Species fanny pack and strapped it around her not-so-narrow waistline. A genetic gift from her father. *Thank God she didn't get his disposition*, Sarah thought.

"Well, have her back in forty-five minutes! Unlike you, Sarah, some of us have important work to get done this afternoon. We can't take four-hour lunches to go shopping or whatever it is you do."

Sarah caught Jimmy Joe's eyes. He shrugged dejectedly. "I'll tell you what, Dick," she said. "While I'm out shopping,' I'll see if I can find a little compassion for you. Maybe on sale at Dillards. God knows, the voters would appreciate it."

Before either one of her parents could land another blow, Jen grabbed her mom gently by the arm and lead her out the door. It swung slowly shut. Sarah felt a cool breeze and smiled. Being near Dick these days was like having a premature hotflash. Not that she personally knew what that was like, but plenty of her friends had them. The only way Sarah had survived sharing a house with him lately was because he spent all his time in hotel rooms while campaigning. On those few nights a month Dick was home, Sarah managed to be away on business.

Inside, Jimmy Joe sighed loudly. The volunteers nervously returned to their various tasks. "Do you think it's wise to keep antagonizing her, Dick? After all, if you make it past the primary…"

"If?" Dick huffed.

"When you make it past the primary, you're going to need her backing more than ever," Jimmy Joe said. He was a practical man. Campaigns demanded money. And even though Dick had seen lots of money in his career as a banker, he had fallen behind in fundraising. There was just a couple of million left in the coffers at this point, and when it came to running a Texas-sized campaign for Governor, a couple million dollars couldn't even grease all the palms that needed greasing.

"I know!" Dick seethed, as he clenched and unclenched his fists like a small child about to have a tantrum in line at the grocery store.

"Get it under control," Jimmy Joe said as if it were the hundredth or the thousandth time the words had passed his thin, chapped lips. "Or one day soon there'll be hell to pay."

Dick nodded just to get Jimmy Joe off his case. "Forget her. What's on my schedule for this afternoon?"

"Well, you missed that breakfast with the mayor of Wichita Falls at eight. He called, pretty pissed. You'll have to smooth that over," Jimmy Joe said as he ticked down a long list.

"Wichita Falls, hell. You handle that."

"Okay. But don't forget the chili cook-off this afternoon at two down at the lawn in front of the Capitol. The Kilgore Rangerettes are in town and it'll be a great photo op." The Rangerettes were the Rockettes of Texas – a definite photo op for every wannabe politician in the Lone Star state.

"You bet your ass, I wouldn't miss that. Rangerettes. Now that'll put me a good mood." A wicked smile crossed Dick's lips.

Jimmy Joe stiffened. "Dick, keep your hands to yourself. More of that kind of trouble we don't need." Jimmy Joe had already spent the morning dealing with the lawyers for Emily Burton, the striking young woman who had charged Dick with sexual harassment at the bank Dick had managed before he threw his hat in the ring. "We've got enough trouble as it is."

Lee Press-On Penises

Rain fell in sheets. Wet flannel sheets. The whole day had a wet flannel feel.

Business at Lone Star Cuts was slower than usual. Maybe it was the rain. Maybe it was the day-before-payday lull. Maybe, despite the weather, everyone was having a good hair day.

Inside the salon, Ima, wearing a short-sleeved maroon velvet dress with an empire waist that made her look like the princess she knew she was, sat at the manicure table, laying out Tarot cards. She was trying to get the fates to help her decide whether to go ahead and lease out her spare bedroom to the guy who showed up last night in response to her ad. So he had a gun rack, that was nothing new. Although Ima had never seen a gun rack on a motorcycle before.

Judy was in her usual one-legged pose, studying her reflection in the mirror. Her newly-dyed reddish-black hair, curled tight on top and cropped short on the sides, nicely set off her mustard-hued sweater and blue skirt. She looked like some kind of wild jungle bird. An extinct wild jungle bird. No bird that colorful could have survived this long.

Then, ever-so-slowly, like the speed at which the burn from a jalapeno pepper finally makes its way to one's taste buds, she unrolled curlers from the overly-blonde, overly-tanned head of sixty-something Maddy Lou Padilla. The tan looked almost fluorescent on a gray day like today. Despite the torrential downpour, Maddy Lou was dressed in tennis whites, with large gold earrings, a heavy gold necklace and four or five gold bangle bracelets. Thanks to her country club membership, both the tan and the tennis were indoor sports for Maddy Lou.

"She claimed her hip gave out when she rushed the net. Sore loser if you ask me," Maddy Lou complained, re-crossing her legs in the other direction, to prevent her circulation from cutting off and leaving another varicose vein in its wake. Her doctor had told her she needed to stop crossing her legs altogether, but Maddy Lou never listened to doctors who were younger than her children. These days, that meant she didn't listen to any of them.

"Are you sure her hip didn't give out?" Judy goaded.

"Well, she did ride off in the ambulance," Maddy Lou shrugged her bony shoulders. "But I wouldn't put it past her. She'd do anything to not concede defeat."

Judy laughed. She had this vision of Maddy Lou and a blue-haired Southern belle pummeling each other at the net, taking care to avoid breaking a nail, gold jewelry clanging, their senior citizen husbands standing on the next court over enjoying the spectacle of the cat-fight and popping a few Viagra.

"Yeah. And she probably didn't even need that oxygen..." Judy began just as the phone behind the front desk rang. Donna was on her break, so Judy laid down the curler and picked up the receiver.

"Lone Star, this is Judy. Nope, Nan's not here right now." She glanced over at Ima at the far end of the front room and they shared a look. "I dunno. Sure, I'll tell her." She hung up the phone and jotted a note on a pink pad. She walked over to Nan's station, taking care not to step on Maddy Lou's racket, and stuck the note to the mirror.

Judy returned to her client and began to remove the last few rollers.

"So, where is the crazy woman?" Maddy Lou asked.

"Which crazy woman do you mean? You know we got a whole shop full of 'em here!" Judy smiled mischievously.

Maddy Lou motioned towards Nan's empty station.

"Believe it or not, she has a lunch date with an accountant." Judy tugged on the last curler. A little too hard. Maddy Lou grunted. "A match made in heaven, if you ask me. She's got seven personalities and he's got none. So on average, they've got three and one-half personalities each."

A giant paper cup blew by the window. Judy gazed out. With that storm out there, the zoo was probably out for today. Maybe she could take the boys next week. They could see a movie today instead. Stevie Ray had been begging her for weeks now to take him to the latest car chase picture, so maybe she'd give in. Judy hated action movies. But the boys refused to go with her to the chick flicks she liked. And as they kept reminding her, they were too old for animated movies, despite the fact that they still watched cartoons at home on TV. Mark had offered to take the boys to blow-things-up, shoot-them-up movies, but he was only free at night and on the weekends. And Judy had better things to do with him then.

Maddy Lou shifted her weight in the chair and looked at her watch. It was a diamond-encrusted bracelet thing with numbers so tiny Judy didn't know how she could read it.

"Please hurry. I have to pick up my tux for the shindig at Lakeway tomorrow night."

"Tux, huh? Aren't you the trendsetter!"

"I've been wearing ball gowns for so long, I thought why not just see what it's like to be the one at the party wearing the pants," Maddy Lou said impishly.

The front door opened. A large gust of wind practically took it off its hinges. The door chimes were just background noise to the wind and rain. Judy looked up and into the eyes of Gail, soaking wet.

“Hi, Judy.”

It took Judy a second to place the smile. "Hey, there, Gail. I’m surprised you recognized me with my bra on."

"Trick of the trade.” Gail said as she shook like a giant tabby cat that had been caught unexpectedly in a downpour.

"You look terrible," Judy said in Gail's general direction.

"Thanks. That's the look I was going for when I got up this morning,” Gail said as he patted her dripping hair.

Judy looked down at Maddy Lou, who was watching Gail in the mirror. "That's Gail. She did my mammogram last week.” She grabbed a can of *Only in Texas Hairspray* and sprayed every inch of Maddy Lou's hair. Several inches of her freckled neck and shoulders got caught in the crossfire.

Maddy Lou had been examining Gail in the mirror. The word mammogram sent a visible shudder through her. "Don't remind me.” She looked at her watch again. "Listen, Judy. I need to hightail it outta here."

Judy combed the last stray hairs into place and removed Maddy Lou's smock. "Okay. There. Perfection."

"Darlin’, there ain’t no such thing. I know because if there was I'd have bought it by now!" Maddy Lou was serious about that.

"What about Blue Bell Cookies & Cream ice cream?"

"Close, I guess.” Maddy Lou admired herself briefly in the mirror, reached into her oversized gold-toned purse, pulled out two wadded up bills and handed them to Judy. They were hundreds. Judy tried to look nonchalant.

"That should cover it. Keep the change." Maddy Lou picked up her tennis racket. "I just hope this holds up in that storm."

Judy looked at her client's new do. "Don't worry, I used the good stuff. You could survive the Titanic."

Maddy Lou ran from the door to her car. She was fast for an old broad, and her hair didn't budge an inch.

Inside the salon, Judy fingered the bills. "Now I can pay both the electric and the water bill!" She looked around. Even Ima was nowhere to be seen. "Oh, what the hell, Margaritas all around!"

"Were those hundreds?" Gail asked.

"You betcha!" Judy tucked one into her bra. "For a sixty-five dollar style. Not bad!"

"Do all your customers tip like that?" Gail continued as unbuttoned her rain coat.

"Only the ones who aren't my friends." Judy smiled at Gail.

"Then sign me up for the friends and friends program." Judy noticed how Gail's eyes completely disappeared when she laughed.

As Judy reorganized her station, Gail plopped down on the couch and skimmed through a Glamour magazine, making faces at the bone-thin models. Gail was a size fourteen and looked great. Everyone said so. At just a squinch shorter than six feet, there was never any way she'd be a size six. She dropped the magazine on the table and looked around the shop.

There were your usual framed photos of beautiful people in hair styles no one would dare to try. Asymmetrical cuts, bangs that hung four inches below the eyes, wild spiky do's that looked like they could gore anyone who got too close. There were six or seven live plants, including one, a ficus, Gail guessed, that had grown too tall for the room, and had bent at the ceiling and started growing sideways. Flexibility. That was the key to survival.

Finally, Judy waved Gail over. "Have a seat. And try not to drip on my shoes."

"Sorry about that." Gail sat down, water starting to pool on the floor around the chair.

"Hey, I was kidding about the shoes. Look at 'em. They're cheap plastic. Probably made from recycled Frisbees or something." Judy clicked her heels together, Dorothy-like, but her shoes were obviously powerless. "Anyway, it's not your fault it's raining cats and dogs." She looked at Gail, with her head tilted to one side. "It's not, is it?"

Gail shook her head vigorously. Water flew around her. Judy ran her fingers through Gail's hair – even wet, it was as fine as silk. "What kind of conditioner do you use?"

"I dunno. Whatever's on sale." Gail didn't believe in extravagance. At least not when it came to herself. Her dogs got only the best money could buy.

"Really? Boy, are you lucky. My hair would feel like straw if I didn't use Paul Mitchell." It was a line, sure, but in Judy's case it was also true. "Of course, it doesn't help that I change my color more often than I change my bra."

Gail laughed and then said in mock horror, "Ooh, and to think I touched those breasts!"

Judy grabbed a smock and placed it around Gail's shoulders. "Now just loosen the smock and pull out your right breast," Judy said. "Oh, wait, that's your line."

Gail laughed heartily.

Judy leaned up against her counter, crossed her arms over her chest and looked at Gail. Even though they'd only met the once, Judy felt Gail was already a friend. Judy was never wrong about these things. Although she was wrong about many other things.

"So, how are you?" Judy asked, obviously nowhere near ready to get to work on Gail's head quite yet.

"Besides soaked?" Gail smiled that magical eye-disappearing smile of hers. "Actually, I've had a great day. All of my patient's x-rays came back negative."

"Mine too!" Judy said emphatically. She'd gotten the results two days ago. A heavy weight – much like Juan when he'd pass out on top of her smelling like Cuervo Gold – had been lifted. Judy had said a silent prayer to her mom when she'd hung up the phone after talking to her doctor on the phone.

"I know. I heard. Great!" Gail wished she could given Judy the good news herself. But radiology techs weren't allowed to do that. That was the doctor's job, although ten out of ten times, Gail could read the x-rays just as well as the radiologist.

Judy could see there was something else Gail wanted to tell her. "You have more good news?"

"How could you tell?" Gail asked.

"Oh, perhaps the Cheshire cat grin on your face. Spill."

Gail paused just for a second to build the suspense. "I sold a short story."

Judy picked up her scissors. "I didn't know you wrote."

"Well you can't learn everything about a girl over a mammogram." Gail smiled.

"That's great," Judy said.

"I don't really write. Just thought I'd give it a try. I sent a story in to a literary magazine called "Tempest." They bought it. It's only $25 and probably no one will read it, but I'm so excited. It's the first time my name has ever been in print. Well, except on my checks." They both laughed. "I'm pretty sure that doesn't count," Gail added.

"Congratulations! Judy finally picked up a comb and looked like she might actually get started on Gail's hair.

"Thanks." Gail gushed as she caught a glance of her wet self in the mirror. "The funny thing is you're the first person I've told. Except Glen, of course. And I hardly know you."

"Oh, I have that effect on people," Judy replied. "You wouldn't believe the things people tell me while they're in my chair. Maybe it's because I oil it down with truth serum every morning."

"Well, that explains it." Gail said as she noticed the picture of Stevie Ray and Nolan on the shelf below the mirror, next to the blow dryer. "Those your boys?"

"Yup!" Judy put down her tools and picked up the photo. She fingered the frame. It was dark blue with hand-painted gold stars and moons. She held it closer so Gail could get a better look. Gail took it and held it further away. At her age, things looked better at arm's length.

"This one' s Stevie Ray and this one's Nolan," Judy boasted.

"Cute!" Lynn agreed with a nod. "For humans, I mean." She laughed.

"Yeah, they're cute most of the time. When they're not trying to kill each other." Judy still had trouble accepting how much her boys fought, even though every mother of boys she knew told her it was normal. Sometimes, she was afraid they'd actually draw blood. But so far, so good. And overall, Stevie Ray and Nolan were good kids. Neither one had yet been pierced or suspended from school. Although she was pretty sure both would happen soon.

"Wanna hand me my purse?" Gail asked motioning towards the couch. Judy retrieved it and handed it to her. Gail pulled out her wallet and produced a picture of her dogs sitting on Santa's lap. "My boys. That's Hans and this one trying to hide his head behind Santa's coat is Bubba."

"Hans and Bubba?"

"Yeah. We were gonna call them Hans and Franz after the old *Saturday Night Live* characters. You know, back when it was funny. But Bubba just did not look like a 'Franz.' He looked like a good ol' boy from West Texas. So we went with Bubba."

"What kind of dogs are they?" Judy asked.

"Dachshunds. Purebred," Gail said with pride.

"They're huge! Purebred, huh? Not my boys, they're mutts!" Although I do know who the father was. But he definitely didn't have papers." Gail returned her photo to her wallet and handed her purse back to Judy, who tucked it under her counter, then picked up the scissors and comb again.

"So, you want to dump this depressing hair style? You look like a Baptist preacher's wife."

"No. Just a trim. Glen likes this style."

"Glen? What is he, a Baptist preacher?"

"Engineer. Presbyterian, actually."

"You always do what your man likes?" Judy cajoled.

"Nah." She smiled. "I'm lucky. Glen is a great guy. I could probably come home bald and he'd say, 'Honey, you look great!' As long as I was standing there naked."

Judy picked up the electric razor and switched it on mockingly. "I could do that!" She turned it off and picked up her scissors instead. "So where'd you meet this wonderful guy?"

"Believe it or not, Luby's cafeteria. He was getting the barbecued chicken, I was reaching for a piece of pecan pie and it was love at first bite."

Judy tapped the scissors against her chest and sighed. "No fair! So, how'd it happen? You turn to the Luby's cafeteria lady and say, 'I'll have this guy with lots of whipped cream?'"

"Close enough."

"All I ever get when I go out to eat is five pounds heaver." Judy studied Gail's hair. "You wanna keep your part on this side?"

"Yeah."

"Okay, just trying to mix things up for you. I know changing a part can be a huge risk!" Judy took her first snip. "So what's the story about?"

"What story?" Gail asked.

"The one you sold, silly. Try to keep up here. I know it's faster-paced than at the mammogram clinic. Plus, no one's naked." Judy mulled over the similarities between a hair cut and mammogram. *At least if you get bad news here,* she thought, *it'll grow out eventually.* A sliver of sunlight cracked through the sky and bounced off Judy's mirror, reminding Judy to let her mind follow the sun, not the clouds. *Maybe the zoo is back on*, she thought. That'd make Nolan happy.

"It's about a dysfunctional family that has an accidental reunion in a mental hospital," Gail said. "I know it sounds kind of depressing, but it's really not. It has a happy ending."

"Dysfunctional family? Mental hospital?" Judy asked. "Have you been spying on my family?"

"No, it's something with which I have lots of experience."

"Don't we all? You, though, I wouldn't have guessed," Judy shrugged.

Gail nodded her head.

"Hey, stop that or you'll be going home with a new trend-setting hair do. We'll call it the 'Little Bit o'Bald.'"

Gail held her head still. "We wouldn't want that." Judy snipped again and Gail continued her story. "My mom was married five times and my dad four times. In our family photos, every other person's head is x'ed out with black Magic Marker."

Judy laughed and kept trimming.

Gail continued. "Then there was my real mom, the alcoholic..."

Judy stopped. "Ditto my dad. He ran a bowling alley. I think because it was the only job where no one would notice he was drunk all the time. Personally though, I think he also had a thing for bowling shoes. Talk about dysfunctional"

"I can beat that!" Gail asserted somewhat proudly. "My dad wanted a boy when I was born. Since I lacked the proper anatomical equipment, he decided to shave my head and pretend I was a boy until I was fourteen."

Judy shook her head. "Oops! Sorry about the electric razor thing. Maybe I should hide that so you don't get any nasty flashbacks." She wasn't surprised at Gail's candor. There really was something about sitting in that chair that made people want to open up and share their lives. Too bad Judy didn't make $150 an hour like the rest of the therapists in town.

The phone rang. "Excuse me. Your dysfunctional family will have to wait a second." Judy pointed toward the phone with her chin. "One of my other dysfunctional clients, probably."

Judy returned less than thirty seconds later. The sun was getting brighter, but the wind was still blowing. "So, your dad wanted a boy, huh. I guess you're lucky he didn't go down to Castle Superstore and pick up some equipment. Could have superglued it on."

Gail laughed. "Duct tape. That was more dad's speed."

"I've got it!" Judy put down her scissors so she could use both hands for emphasis. Donna walked in the front door, carrying an armload of packages. "Lee Press-On Penises," Judy said with a flourish.

Donna dropped her packages on the couch. "What's this about 'Lee Press-On Penises?'"

"Wouldn't you like to know?" Judy chided. "If you're going to be out, you're going to miss all the important stuff. Gail, this is Donna. Donna, Gail. I told you about her. She did my mammogram."

"Hi, Gail," Donna said.

"Hi, Donna, nice to meet you," Gail said smiling.

Donna gathered up her purchases and sauntered to the back room to store them in her locker.

Judy looked in her general direction, thinking about God knows what. "So, what were we talking about?"

"My dad." A serious expression crossed Gail's face, but quickly passed. "You know, it really pissed dad off when I got breasts. My therapist says maybe that's one of the reasons I went into radiology. It's like I have this real respect for the power of breasts because mine were the ultimate revenge on my dad."

"Hey, I feel at a disadvantage here. I showed you mine." Not looking at her work, Judy accidentally caught her scissors in Gail's dangling earring. "Sorry. Maybe we should take these off. I don't want to rip your ear off and have to call 911 again."

Gail removed her earrings. Silver dog silhouettes. Now that they were off, Judy could tell they were those hot dogs, Dachshunds.

As Donna sat down at the front desk, lunch break now over, Gail stared at her reflection in the mirror. "Do you think I'd look better with bangs?"

"No. You'd look like the star of *Gidget Has a Mid-Life Crisis*. That'd be the one where she takes her surfboard and beans the guys on the beach and then heads off to Paris, France not Texas, to a health spa, where a buff guy named Sven oils her up. I think I should take a little more off."

"No, I like it the way it is now." Gail squinted to get a better look at herself. "Gidget, huh? And what makes you think I'm middle-aged?"

"You told me you were over forty when we first met."

Gail looked perplexed. "I did? I don't remember."

"Of course not. You were ogling my fabulous tiny breasts. Not to mention you ARE middle-aged. I hear THOSE people tend to forget things."

The door opened again. A good-looking guy in his late twenties, wearing faded blue jeans, Birkenstocks, and a Lyle Lovett t-shirt, ambled in.

Donna looked up. "May I help you?"

The good-looking guy blinked, his long eyelashes temporarily darkening his denim-blue eyes. "I'm just waiting for Ima."

Judy, who had been giving Buster – that was his name if she remembered correctly – the once-over ever since he arrived, noticeably straightened her posture. "You must be the infamous Buster. We've heard all about you." She winked at Gail.

"Is she here?"

Donna pointed towards the back. Judy chimed in, "She's either cleaning up her manicure station or channeling Joan of Arc, I forget which."

Buster walked off towards the back. All eyes on the Levis 501, Waist 31, Inseam 34 tag on his rear pants pocket. Or somewhere in the near vicinity.

"Whew!" Judy, Gail, and Donna exclaimed simultaneously.

Judy popped Gail on the shoulder with her brush. "Hey, you're not allowed to look, you're married."

"Oh honey, I always look!" Gail said with a giggle. "It's like art or expensive shoes. You look and get it out of your system."

"Well, maybe that works for old married folks," Judy said, brushing away a stray hair from Gail's nose, "But I look and I get all revved up. I have to find a completely different way to get it out of my system." *Thank God, Mark is coming over tonight*, she thought. There's nothing worse than being all sexed up with no place to go.

"Actually I shouldn't look either. Because I've got Mark."

"Mark," Gail asked. "Why is this the first time I've heard anything about him?"

"Well, I didn't want to spill all my secrets right away. He's this guy I've been dating. Not my usual type, but it's working out so far." Judy put down her comb and picked up her blow dryer. "Speaking of my usual type, which, by the way is the bad boy/jerk type, why don't you tell me your big secret about Dick Sloan?"

A dark cloud, like the ones quickly dissipating on the horizon, crossed Gail's porcelain brow. "What secret?"

"Oh, come on," Judy harrumphed. "I mentioned his name the other day over my naked boobies and you went all psycho on me."

"Psycho's such an ugly word," Gail attempted to joke, but her smile had disappeared and the cloud returned. "I hate to ruin such a lovely day."

"Ah, come on. I'm still so pissed at what he did to Nolan. To think he could be our next Governor. It's enough to make me want to vote! What did he do to you?"

Gail looked down at her ragged nails. Her voice lost its volume and spunk. "He date raped me in high school," she whispered.

"What?" Judy whispered back.

Gail stared at the floor. "We were in the same class in high school. He was the quarterback on the football team. Which I could never figure out because his whole family had these short stubby fingers." Gail paused for a second. "I was this tall, real skinny nerd. The joke around campus was that I wore a pocket protector on my nightgown."

"That's pretty nerdy all right." Judy said, trying to lighten the mood.

"So, when Dick Sloan asked me out to the movies, I couldn't believe it. I'd only had one date before and my mom had set that one up." Gail paused again. "It turned out, the only reason he asked me out was that his shit-kicker friends bet him he couldn't get me to go all the way." Gail stopped. Judy stood, rapt, holding the blow dryer for no reason at all.

"Did he?" Judy asked sucking in her breath..

"He tried. We went to the movies. Saw *Raiders of the Lost Ark.* Then, on the ride home, before I knew what was happening, he ripped off my blouse and was trying to unzip my jeans. But he couldn't. I don't know if it was his stubby little fingers or me smacking him with my purse. He kept yelling at me, 'You bitch, you know you want it!' I cried the whole time."

"That must have been terrible." Judy hadn't made so much as a snip on Gail's hair. She was in her therapist role now, not her hairdresser role.

"It got worse. The next day, I overheard some cheerleaders talking about the bet. It was humiliating. The only thing I regret is not going to the cops that night and reporting it. But I didn't think anyone would believe me. After all, I was just a nerdy girl and he was the quarterback."

A few seconds of silence passed, which Judy interrupted by picking up a pair of sharp scissors. "What a jerk! I say we snip him. You get him for date rape first. Then I'll get him for attempted kidslaughter."

"I'll pass," Gail said, the cloud lifting from her face. "He's not worth it. Besides, with his wife's money, he could afford the lawyers to send me to prison if I decided to neuter him."

"Oh, darn! I was so looking forward to fixing him. Well, I guess I should use these scissors for their intended purpose." Judy snipped some stray hairs. "Anyway, from what I can tell, Sarah Sloan isn't too fond of hubby either. You should have heard them shouting at each other in the parking lot a few weeks ago. In fact, that was the day Dick ran over Nolan. She took it really hard."

Gail wasn't really listening. She was still thinking about high school. She shook her head as if to shake off the memory. "I'm over it. Three years of therapy and I'm as normal as they get."

"Yeah, you're normal all right. Feeling women up all day. Dreaming of castrating politicians. Daring to wear this out-of-style hair do."

"I just stick with the classics." Gail said proudly.

"Honey," Judy drawled exaggeratedly, "this hasn't been a classic since the seventies."

"I like it. It's kind of 'That Girl.' Do you know that old show with Marlo Thomas?," Gail began to sing. "Diamonds, daisies, snowflakes, that girl."

"Oh, you mean that girl, as in 'Who's That Girl with the Weird Hair?'"

Gail laughed. "Have you ever thought about doing stand-up?"

"All the time." Judy's eyes lit up. "You think I live to cut hair? Once the boys are old enough to…" She paused. Old enough to what? Live without her? Go on the road? Have girlfriends who cook for them? Have boyfriends who cook for them? Judy wasn't sure exactly what she was waiting for.

Gail shook her head. "Why not give it a shot now? Here I've been talking about the worst night of my life and you managed to make me laugh."

"Well, it is my job around here. You think they hire me to cut hair? A toothless orangutan could have gummed your hair off faster than I cut it."

Gail laughed again. "Glen and I go out once in a while to this place called Just Fer Laffs. We were there a few weeks ago. They're having this contest to find the funniest people in Austin. The winner of each round gets $500."

"$500?" Judy's ears perked up, like a dog who hears the word "Cookie" while digging holes in the yard.

"Yeah. A guy from Glen's office tried it. He sucked!"

"Oh, that's the way to encourage me." Judy reached for the $100 bills in her bra. She unfolded them and looked at them reverently. "$500? With $500, I could finally fix that hole in my floor."

"You have a hole in your floor?"

"Yeah. Stevie Ray almost broke his leg in it the other day. And one of the cats has been missing for a few days. He has been replaced by a family of raccoons, so we haven't really missed him."

"You should do it. Stand-up, I mean," Gail said. "This town needs some more female comics. I'm tired of hearing about men and their penises."

"You talking about comedy or politics?"

"Both."

Judy's eyes narrowed and she looked at the photo of her sons. "$500? I'll think about it." Five hundred dollars might be just enough to make her serious about comedy.

"You let me know if you decide to go for it. Glen and I will come for moral support."

"I might do it just so I can meet this perfect man of yours." Judy went to front desk and tore a page off the *While You Were Out These Idiots Called* notepad. She jotted down the words 'Just for Laffs.' "I could invite a bunch of friends and threaten them if they don't laugh hard enough to make me win."

Energized, Judy finished Gail's hair in just fifteen minutes. By then, even the sidewalks were dry. And no one was thinking about Dick Sloan and his stubby little fingers.

No News is Good News

It seemed that no matter what she did (and she didn't do much), the inside of Judy's trailer always looked like a twister had touched down inside. Most days, she didn't really care. When you're raising two young boys, if cleanliness and order are your priorities, you're in for many, many years of heartbreak and disappointment.

Judy sipped her coffee – decaf, not high octane. Now that she was having regular sex, she didn't feel as much need for leaded.

She sorted through the mail. Phone bill. Electric bill. Outrageous garbage bill. It was incredible how much garbage two boys could produce. She could fill up the trash can just emptying their pockets. Lots of bills. No child support payment from Clay. As usual, he'd probably be three weeks late again.

$500 would come in real handy, she mused as she thought about the comedy contest. Judy had already called the club, which spelled its name, *Jest for Laffs*, just for laughs, and gotten the details. The next contest was next Saturday. She had written down the info on the back of the water bill envelope, which was hanging from the freezer door by a Pizza Hut magnet. And she had tentatively lined up a babysitter. But she wasn't sure she was going through with it.

The phone rang and Judy lazily walked across the kitchen to answer it. Judy's phone still had a cord on it. She'd tried a cordless phone, but she could never find it.

At least with this old phone she could always follow the umbilical cord.

"Hello," she answered lethargically.

"Hey beautiful, what's up?" It was Mark.

"Hey yourself. Well, me, but just barely. You know I like to sleep in when I get a chance."

"So I've been told. One day, I hope to see it for myself." Mark was hopeful, but not pushy.

"Oh, you must be a glutton for punishment. Me without make-up. That'll make you turn and run." Even though they'd had plenty of sex, Judy still felt it was wrong for him to sleep over. *What kind of message did that give her boys*? She wasn't sure if she was more afraid they'd think their mom was slutty or about to get married again.

"Hey, you can't shake me that easy," Mark teased. "I called because I was wondering if you'd do me the honor of being my date to my company party the Saturday after next?"

"That would be great." Suddenly, Judy remembered the comedy competition. "Wait. Saturday after next? I might have plans."

An explosion came from the living room, which was officially located, according to the guy who sold her the trailer, four inches from the permanent indentation her kitchen chair had made in the cheap vinyl kitchen floor. Judy looked up from the bills and over at the TV, where Nolan and Stevie Ray were lying, stomachs down, noses practically on the set. They were so quiet, she almost hated to ask. But ask she must. After all, it was her job. Haircutting and comedy, those were just her hobbies. Or bad habits.

"Just a second, Mark." She put her hand over the receiver. "What are you two watching? " she grilled.

"*COPS*," Stevie Ray admitted hesitantly, without turning around to meet her eyes. *COPS* was one of the shows she had explicitly forbidden them from watching, in part because she was sure their father would show up on it one day. And not in a uniform.

She held out her hand and Stevie Ray handed her the remote. Off it went. "What have I told you about shows like that?"

"That if we watch them, we'll end up in prison doing laundry for guys who call us 'Sugar Buns?'" Stevie Ray mimicked Judy so well it drove her crazy.

"Yes. That. Besides, these shows give your brother nightmares."

"Do not," Nolan rolled over and insisted.

"Do too," Judy shot back.

"Do not."

"Well, they give me nightmares."

"Mom!" Nolan whined, staring pathetically at the screen as if it staring alone could make it come back on.

"You know the rules," Judy said, trying to act like a stern disciplinarian, something she was not, "Every time I catch you watching something you're not supposed to, it's no TV for the rest of the day."

At that, Stevie Ray jumped to his feet and headed toward the door. "I'm going to go ride my bike."

"Me too," Nolan sing-songed as he bounded after his brother. He almost fell into the hole in the floor. Stevie Ray made that face that big brothers are required to make when they realize their younger brothers are going to follow them around no matter what they do. "Mom!" he whimpered, to no avail.

"You two be careful and wear your helmets. On your heads this time. With the strap done up. And be home in an hour." The boys were already out the door and down the steps before she finished her sentence.

Finally, Judy returned to the phone. “Remind me not to have any more children,” she said to Mark.

“Okay,” he laughed a little nervously. “So this thing you might have next Saturday night? You going to tell me about it or am I going to have to worry that you’re seeing someone else?”

“Oh no, it’s not that!” Judy hesitantly told him her tentative plans. “I’m not sure I’ll do it. I’m not ready.”

“Sure you are. I’m sure you’ll win.” That was Mark. Always saying the right thing. “I’m just sorry I won’t be able to be there. I’ve got to go to this party.”

“That’s okay,” Judy said a little relieved. “This way I can talk about you on stage.” She laughed. “I’m sorry I can’t go to the party, though. Maybe next time?”

“Absolutely. But I will see you Thursday night, right?” He asked.

“Yep. Can’t wait.”

After hanging up the phone, Judy sank into the couch, a monstrosity the color of mildew she'd found one day outside a frat house at the university. Even though she’d steam-cleaned it and Scotchguarded it and performed an exorcism over it, it still smelled like beer and bong water. She thought about Mark, then she thought about the comedy competition. It was all too much, so she clicked the TV back on and changed channels to the news.

The anchor man with the bad wig was on.

"According to the latest Channel 2 News poll, Dick Sloan leads his fellow Republican Rich Clemson by twelve points. We go now to Hector Alvarez at Sloan headquarters…"

“Shit!” she shouted as she turned off the TV. Good thing the boys were outside, or else she’d have to put a quarter in the cursing jar.

****

Lynn's head was pounding and had been all day, despite six Ibuprofen. She didn't know if it was allergies or barometric pressure or that constant nagging fear of losing her job. It didn't really matter. A pounding headache is a pounding headache.

Still dressed in the taupe suit she'd put on for work fourteen hours ago, Lynn sat on the sofa and rubbed her temples with a eucalyptus/lavender cream. The sofa was from Spiegel's catalog. The cream was from Sundance. The headache was from Life. Page 118. Also known as "This isn't what I had in mind."

"Hey, Coco!" she said as she looked up and saw the cat gracefully traverse the living room towards her. Coco was as black as a solar eclipse. Lynn's furniture and carpet were white. Needless to say, her Hoover worked overtime.

Coco slithered from the floor to end table to the sofa without disturbing a knick-knack. Lynn didn't have knick-knacks. They just needed dusting. There was just one picture on the fireplace mantle. Judy and Lynn, with Stevie Ray and Nolan, at Six Flags two years ago, standing in front of the log ride, with water splashing up behind them. They should do that again. If she could find time to get away.

Lynn picked up Coco and breathed him in, then turned the TV on. She'd TIVOed the news, so she hit play and Dick Sloan's smiling face appeared.

"It's clear from the overwhelming lead I have over my opponent, that the voters of this state want a return to our conservative roots," he said grinning at the cameras. *Apparently a little sex scandal only helps a campaign in this state,* Lynn thought with disgust. "A return to family values. When I am elected Governor, we're going back to the high moral standards this state used to be known for.

One of the first things I'll do is make sure that a bill prohibiting same sex marriages passes."

Lynn hit the "Stop" button with her index finger so hard she almost jammed the knuckle. Too bad she couldn't just press "Stop" on Dick Sloan, too. The asshole.

She looked down at Coco. "Well, that settles it. If he wins the election, we'll just have to move." Coco, unaffected by the news, started to snore peacefully in her lap. "At least he didn't say anything about human-cat couples," Lynn said looking lovingly at the pile of fur on her lap. "And that's the only couple I'm a part of anyway."

She stroked his ears gently. "How would you feel about moving to Denver, Coco?"

****

"Where are all the bowls?" Glen asked his wife.

"I got the last clean one. You'll have to rinse one out." Gail looked up at her husband with a smile. "I guess someone forgot his turn at dishes last night."

"As I recall, we were a little busy doing other things," Glen winked at her. Gail took another bite of mint chocolate chip ice cream.

"Yeah, that was fun. We should make love in the laundry room more often."

"Only if we get insurance that covers a chiropractor," Glen said as he rubbed sore spot on his knee. He straightened up, walked to the kitchen sink and ran hot water into a blue bowl with a white stripe around the top edge. Glen scooped what was left of the ice cream out of the carton into his bowl and sat down across the table from his wife.

"Did I tell you about the funny woman I met a couple of weeks ago? A hair stylist? She cut my hair today."

"You got your hair cut?"

Gail stared at him with that "I can't believe you didn't notice" look that women use even after they've been married for years and know better. She kicked her spouse gently under the table. "Anyway, I'm trying to get her to enter the contest at Just Fer Laffs. Maybe we could go see her."

"Sure. She's gotta do better than Ed."

"That's what I told her." Gail looked toward the living room where Hans and Bubba were sleeping on the blue chenille La-Z-Boy sofa, the foot rests extended so they could stretch out to their full lengths.

"Hans! Bubba! Ice cream!" Gail called to the dogs.

"Did you get all the chocolate chips out?" Glen asked, attentive dog father that he was.

"Yep," she said as she put her bowl on the floor. Just then the Dachshunds ran into the kitchen, shoving each other with their snouts to be first to get the treat. The bowl was licked clean in a nanosecond.

As Glen finished the last of his ice cream, the dogs waited patiently, their tails wagging like they'd just invented squirrels and wanted to relay the message via tailegraph. Gail got up and rinsed out her bowl. Not that there was anything to rinse out.

"Hey, I'll do the dishes. It is still my turn," Glen said.

"Okay. I'll go watch TV." Gail wandered into the living room. She lowered the foot rest on one side of the sofa and settled in. Almost immediately, Hans and Bubba came running in behind her, licking green ice cream from their muzzles. They ran up the ramp to the couch. Gail had finally talked Glen into making the ramp last year when Bubba was having trouble with his back. She had attached the side rails so the boys' feet wouldn't slip off on the way up. Hans jumped into her lap, while Bubba tried to rub the ice cream from his snout onto the sofa. Gail turned on the TV.

"The question isn't who will win the Republican primary, but by how much of margin Dick Sloan will win it and whether that will be enough votes to translate into a win in the November race." The guy with the bad toupee looked overly happy to be reporting this.

"Asshole!" Gail spat at the television.

"What?" Glen inquired from the kitchen.

"Nothing, honey. Just come sit with me and the boys." She clicked the TV off.

Join the Fan Club

Buzz, buzz, whack. Buzz, buzz, whack. Every four seconds, a small clip-on fan attached to the wall separating Lynn's cubicle from Ron's banged into the metal coat hook. It was annoying the hell out of Lynn. *If he'd just lose a couple pounds, then maybe he wouldn't be so hot all the time,* she thought, as she hovered over her keyboard, trying to come up with the next line of her story.

Finally irritated into action, she got up and peered around the half-wall. Ron was probably out, hanging around the receptionist's desk gobbling down the M&M's she kept in a glass jar on her desk. Lynn detached the fan, hid it under her Donna Karan jacket and stashed it away in the cabinet behind the Cremora and coffee stirrers in the break room. He'd find it. He always did. But for a few moments at least, she'd have some peace and quiet.

She sauntered nonchalantly back towards her office, a faint sense of accomplishment following her. Already she could feel her headache dissipating. She hated having a headache on a good day. And today had started off so good. *Snowbunny,* whose real name was Annie, wanted to meet her in person. It was the kind of news they don't print in newspapers, but the kind that made headlines in her heart.

Lynn broke out in a smile when she returned to her cubicle. She tried to get back down to business again, but the phone interrupted her scattered thoughts. Shoving aside a half-eaten Subway sandwich, Lynn picked it up quickly. "Lynn Larson," she said, the smile carrying over into her voice. She could see from her Caller ID it was Lone Star Cuts.

"Hi, Judy!" Lynn asked almost perkily.

"Hi! Listen, I have good news," Judy said immediately.

You and me both, Lynn thought.

"Yeah? Great! Spill." Lynn's fingers were still perched on the keyboard, but she wasn't thinking about the story on the Texas prison system, not when 98% of her body was wondering whether she could afford tickets to Denver and what she would wear once she got there.

Judy rambled on and on about some woman named Gail she met during her mammogram. *Only Judy could make a friend over a mammogram,* Lynn thought.

"I'm going to do it. At Jest Fer Laffs. On Sixth Street. So, can you make it next Saturday night?" Judy asked, finally stopping to take a breath.

Lynn lassoed her brain from its errant path. Judy was finally going up on stage. Good for her. "Of course. You know me, I don't have any plans." *At least not right away,* she thought with a smile. "As long as I get Coco fed on time, so he doesn't turn me in to the SPCA."

"Do you think I can lose fifteen pounds by next weekend?" Judy asked with a laugh. Then she remembered a conversation is supposed to have two sides. "So what's up with you today? You sound very chipper. Did you get another big story?"

Lynn paused as she glanced around to see if anyone was near her cubicle. No one. "Better. I was going to wait and tell you this on our walk, but, I've met someone. Kind of."

"That's great! Where? The YWCA?" Judy asked.

"Funny!" Lynn giggled. The thought of the YWCA as a lesbian pick-up place was incredibly odd. And funny. "No, I met her on the Internet. In a chat room."

"So what does she look like?"

"Cute. Freckled. Adorable. She's from Denver, but she travels a lot. Her name is Annie. She sells medical equipment. And no, just in case you were gonna ask, I don't think she's a drug dealer."

"Well that's just great! 'Bout damned time. People were starting to talk about you and Coco. After all, he is a he, you know, even though you had him fixed." Judy said. She paused and Lynn could hear Donna talking in the background. "Uh-oh. I've gotta go. There's someone in my chair actually expecting a haircut."

"OK, I'll see you Saturday," Lynn said.

"Promise you'll laugh."

"Of course I'll laugh."

"And if Annie happens to be in town for a drug deal, bring her too!"

Judy hung up. Lynn sat for several seconds listening to the dial tone.

She finally put the receiver back on the cradle and swiveled around to face her computer. Suddenly, Lynn signed off her word processing program and got online. A big grin spread across her face as she clicked onto Gaydate.com. She didn't even notice that the fan had been replaced. Buzz, buzz, whack.

Nerves of Caffeine

Tears welled up in Donna's eyes. Big allergic crocodile tears. They ran down the bridge of her nose and onto the weekly schedule below. Donna was allergic to half the stuff in the salon. It was one of the hazards of being a receptionist /bookkeeper/diplomat at a beauty salon. Along with not being able to go to the bathroom enough. Kleenex and Depends, that's what you needed to do her job.

Judy ambled in through the front door. Donna glanced at the clock. It was only 9:15. "Why are you so early?" she asked sniffling. "Your first appointment is not until 10:30."

"I couldn't sleep. The boys are with The Evil One..." Judy halted.

"Don't tell me you're nervous about tonight?"

"A little." Judy said, under her breath, as she tapped her fingernails on the counter.

"I don't see why. You're on stage here every day."

"This is different," Judy said looking around.

"How?" Donna asked sniffing. "The whole audience is going to be mostly clients and customers. The only difference is you won't be screwing up their hair while you're doing your bit."

"Set," Judy replied emphatically.

"What?" Donna asked.

"It's called a set. I learned that from a book I checked out from the library last week," she said with pride.

"Well aren't you the consummate professional?" Donna was impressed.

"Professional, yes. But am I funny?"

Donna blew her nose. Judy walked off toward the shampoo stations, grabbed a *People* magazine and paged through it, trying to quell her queasy stomach at least until her next client arrived.

Everyone's a Comedian

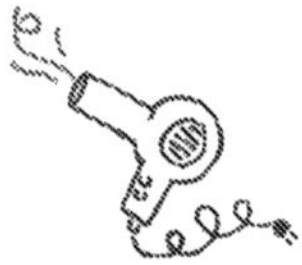

Judy stood in the dark hall just off-stage, wearing a long purple shirt, jeans, a brand new pair of 3" ankle-high black boots from Payless, beaded earrings, and purple nail polish. She felt nervous. She felt nauseous. She felt her stomach to make sure it was still located mid-torso and not somewhere up near her throat.

Some guy was up on stage, doing his set. People were laughing. Down the hall, close to the bar, two other comics, both male, were pacing. She'd occasionally hear part of their sets as they talked to themselves and checked their 3 x 5 cards for their next joke. Judy's set was written on a napkin she had tucked in her jean pocket. She peered at the guys through the darkness, sure that neither had spent and hour and a half trying to decide what to wear. In fact, from the looks of one of them, he'd been sleeping in his clothes for weeks. The other looked like he'd been involved in some horrible accident in a sporting goods store – everything he was wearing, from his baseball cap to his Nike Air Jordans, was stamped with a logo.

Logoless, Judy peeked out at the crowd, peering around the red curtain that smelled like a hundred years of smoke despite the fact that Austin bars had been smoke-free for a couple of years now. *Thank God for that*, Judy thought. She hated the smell of cigarette smoke. Not to mention the consequences. The curtain was also covered with weird stains, looking like hundreds of people had used it as either a napkin or a sweat towel in the past twenty years.

Thanks in large part to Judy's one-woman campaign to pack the audience with friends and clients, the place was nearly full. There were about seventy-five people sitting at tables and another fifteen or so on red barstools that were lined up along the back wall under the photos of comics, most of whom were goofy-looking guys Judy had never heard of. They looked like they hadn't changed all that much since junior high school. Same goofy look. Same goofy clothes. *They probably still have condoms in their wallet from the eighth grade*, Judy thought with a smile.

Judy knew most of the people in the audience. *I'm not sure if that's a good thing or a bad thing*, she thought as she held onto the curtain like Linus holding on to his security blanket. Then, realizing what she was holding onto, she dropped it in disgust. *They know most of my stuff already.*

Gail sat at a front table with Glen beside her. Judy thought he was cute. Lynn and Donna were at the table next to them. April and Leon were right behind, April glancing towards the door every few minutes. Bruce was late again. Leon was wearing his orange cowboy hat. Was it any wonder the man couldn't get any woman to date him twice? Too bad he wasn't gay – he'd have had better luck. Not far from April and Leon, was a new client Judy had seen for the first time Friday, with some other people she didn't know. They looked kind of familiar, too, but she couldn't place them. Bernadette was nowhere to be seen. Probably outside smoking. Bernadette's smoke-free days had ended on week four. Ima had said she couldn't come because her biorhythms were all wrong.

Then wistfully, although she knew he couldn't be here because of his party, Judy looked around for Mark. Sadly, she had to accept the fact that he wouldn't be in the audience tonight. She felt a pang somewhere near her heart. It was either nerves getting the better of her, heartburn, or – could it be – love?

But the show must go on, so Judy took a deep breath and suddenly felt calm. *Maybe Donna was right*, she thought. Maybe joking with clients every day is kind of like being on stage.

As she mentally ran over the one new jokes she'd written on the way to the club, she could hear the emcee walking back on stage. She braced herself.

"Our next comic works as a hairstylist here in town. Her friends think she's very funny. Let's see if you agree. Ladies and gentlemen, Ms. Judy Parker!"

Judy grabbed her glass of water and a pair of scissors from where she had set them down on the stack of liquor boxes and walked out to the microphone. The audience clapped and chanted, "Judy, Judy!" She could make out Lynn's voice above the rest. She put the water on the stool next to the mike stand and then took the microphone out of the stand.

Finally, she looked out into the audience. The lights were so bright, she couldn't really make out anyone's face, including the two new faces that had just come through the door as she was walking onstage. Bernadette. And Mark. He had left the party early, despite the potential consequences to his career. He wouldn't miss this for the world.

"So, how are y'all doing tonight? Me, I'm a little nervous. But, not to worry. I eat at Taco Bell all the time, so I'm used to being queasy." She smiled and told herself to relax.

"As our lovely emcee, Dave, said, I cut hair for a living." She snipped at the air with the scissors, then put them down on the stool beside her glass of water. "I work for Lone Star Cuts. Our motto is 'If they're busy staring at your big hair, they'll never notice your big thighs!'

People were laughing. "We get mostly women clients at Lone Star Cuts. Is it just me, or are there some weird women in this city? The other day I had this 82-year old in my chair. Her natural color is blue. She wanted me to corn row her hair. So, I said, sure, Mom. Whatever you want."

"The most popular style down at our salon is the double-wide. It's so big we have to pump it full of helium just so those women can hold their heads up. You've seen those bobble-head dolls? We invented those. Real women walking around trying to keep their big hair up. That's why we came up with the helium idea. That and everyone thinks I'm so much funnier when there's helium wafting through the salon."

Donna snorted in the front row.

"Being a stylist is a lot like being a therapist. Except instead of a couch, we have a hydraulic chair. And fewer people cry. Well, they cry when I cut their hair, but hey, you gotta take the good with the bad. Besides, if I made anywhere near what a therapist makes, maybe I'd be a little more careful with the scissors."

It was going great. Giggles and guffaws all around.

"One of the best parts of my job is that I overhear the most interesting things. Like a few weeks ago, I'm standing there minding my own business, barely even straining to hear the conversation going on next to me, when I hear one of our clients say her husband is impotent, which is kind of funny, considering her husband's name is Dick."

Gail almost choked on an ice cube. Glen put his hand on top of hers.

"Dick Sloan. You may have heard of him. He's running for Governor. Well, the good news is, if he gets elected, we know for sure it'll be Big Dick, not Little

Dick who takes office. Little Dick, according to his wife, is MIA. Missing or Inactive."

Lynn guffawed. *She has balls, she thought*, chewing the cherry from her Amaretto sour.

"Actually I think we should have a Governor who has to rely on the brain in his head, rather than the one in his pants. Now, there's more to this story. Not only is Big Dick a little dick or vice-versa, he also practically ran over one of my boys in the parking lot in front of Lone Star Cuts. That's where I work. My boss said she'd pay me extra if I mentioned it in my act. Anyway, there's Dick in his big-assed car running down my boy on his bike. Maybe his stick shift doesn't work either."

Judy stopped for a second and took a drink. A few drops dribbled down her blouse. "Fortunately, you don't have to take a coordination test to do stand-up," she ad-libbed, as much to herself as to the audience. "Or a pee test. My blood chocolate level is through the roof tonight."

"I have two boys. It's kind of tough raising two boys by yourself. The main problem is—they have penises and I do not. A few months back, my younger boy had his first erection and he ran into the living room and yelled, 'Mom, what is this? Make it go away!' I explained it to him and he changed his mind. 'Let's call Grandma and tell her!'"

"It's especially tough raising boys in Texas. I feel like I have to teach them to be macho so they can hold their heads, both of them, up high. When they were little, I bought them Tonka trucks with gun racks in the back. You can't say I'm not trying."

Judy took another drink of water, this time without spilling any on herself.

"So, I've recently lost a lot of weight." She waited for the audience to hoot and holler, and they did. "I've been on the Gulf Beach Diet. That's where you go try on bathing suits until the fat gets scared and just drops off. All that screaming you hear in dressing rooms? Fat running for its life."

Donna laughed even though she'd heard that joke a hundred times before.

"I've been exercising mostly and watching what I eat. I refuse to go on one of those liquid diets. They're like oral sex. Not very satisfying and you always end up with a funny taste in your mouth."

Someone at the back of the club shrieked with laughter. Judy was just about to go on to her last two jokes when she remembered something she read: "Always end your set on a big laugh."

"I'm Judy Parker and that's my set. Thanks, you've been a great audience so I guess I won't have to use these." She grabbed the scissors off the stool. Then she put the microphone back into the stand and made a small bow. As she was leaving she could see eight or nine people – Gail, Glen, Donna, Lynn, April, Bernadette and a few others she couldn't really make out – standing and cheering. As she reached the dark hallway once again, her knees began to shake.

****

"And Austin's Funniest New Comic is…" The emcee paused for effect. "The person who, along with our previous two winners will represent our town in the statewide competition in Houston, is… Do you wanna know?" The crowd shouted yes, yes. Judy, standing on stage with the seven other comics thought to herself, *No, we're just standing up here because sweating under hot lights is a quick way to take off pounds.*

"Judy Parker!" the emcee announced, and the room erupted into applause and cheers. Judy smiled, stepped forward and took the $500 check the emcee handed her, folded it and tucked it into her bra.

****

Soon, she was sitting at one of the front tables with her friends who were patting her on the back and telling her how funny she was.

"Con-grat-u-lashuns," Leon slurred, his fifth long neck having taken its effect on his speech. He put his arm on her shoulder either to reinforce his message or to hold himself up.

"HBO will be calling any day!" Lynn said with a smile. "Oh no, I just thought of something. If you get famous, who will I get to screw up my bangs?" Lynn asked in mock horror.

"There's always Nan," Judy retorted, looking her best friend in the eye. Lynn shook her head violently as if Judy had just offered to set her up on a blind date with Leon or something.

Gail interrupted. "Judy, I haven't had a chance to introduce my husband, Glen." Judy reached out her hand, but Glen embraced her warmly. "Great job! Gail's right, you are funny! You should do this professionally."

Donna made a face of mock horror. "Now don't go making her head swell up!"

"Especially now that I don't have the boobs to counterbalance the weight," Judy said. She reached inside her bra and retrieved the check and felt it to make sure it was real. "Although now I can afford.. well.. maybe a boobette."

"You were great! I especially liked the Dick Sloan stuff," Donna added. "Too bad Sarah wasn't here. She would have laughed her big hair right off."

Gail smiled. "Too bad Dick wasn't here. I'd have given anything to see the look on his face."

"You were okay with that?" Judy asked.

"Hell, it was funny and if it keeps us from running after him with scissors and ending up in prison, it's a good thing." Glen squeezed Gail's hand.

Judy looked at Lynn's watch. "I've gotta get home. I don't want to waste this whole check on the sitter. Thanks everyone, for coming to see me. For laughing. For not bringing your constipated friends. I really appreciate it." She re-deposited the check in the Bank of Judy's Bra and stood up.

Gail disentangled herself from her spouse and threw her arms around Judy. "We'll come every time. We're your official groupies."

Judy left the club had almost reached her car when she noticed someone standing beside it. She could barely make him out in the streetlight, but when she heard him call her name, she knew. Mark.

He walked towards her and took her in his arms. She sank into him, all the emotions of the evening flooding over her – the nervousness, the excitement, the joy, and now this. "You came?"

"Of course," he lifted her face toward him and kissed her gently. She noticed how soft his hand was on her neck.

"But the party."

"I cut out early." They just stood there embracing in the parking lot. "You were fabulous. I wouldn't have missed your big night!"

Judy broke out of the hug and mock-punched him. "Why didn't you tell me you were going to come?"

"I didn't want you be more nervous than you already were." He answered, gazing into her eyes.

"That's sweet!" *Hey, whose life is this?* Judy found herself wondering. A comedy contest under her belt, $500, and a great guy? It was almost too good to be true.

Is This Thing On?

"Is that really all you're going to have?" Lynn asked Judy, who was trying to spear a crouton from her paltry salad.

"Hey, not all of us can eat our weight in, what are those?" Judy asked, pointing her fuchsia-tipped finger toward what was now mostly just a grease-stained piece of wax paper with a few crumbs.

"Curly fries," Lynn answered guiltlessly.

"Like I said, not all of us can eat our weight in curly fries and look like you. Some of us actually have to watch what we eat. Even if we DO exercise practically every weekend!” Judy stopped and watched Lynn take her last bite of cheeseburger. "Especially now that I have someone who sees me naked. I mean besides Abbott and Costello.”

“Hey, just because you’re having sex doesn’t mean you have to flaunt it!” Lynn said as she tossed a curly fry at her friend. “Some of us are still in the courting phase.”

“That’s just so great! Anything new to share?” Judy asked.

Lynn popped the last fry into her mouth, took a sip of grape soda, and wiped her hands painstakingly on her napkin until not one subatomic particle of grease remained behind. “Not today. Today is about you.” She reached into her purse and pulled out a tube of lipstick and her Sony tape recorder. She set the recorder on the table and applied the lipstick, perfectly, without checking her reflection. "You ready to get started?"

"I guess."

"This will make a great Person of the Week story." Lynn pushed Erase to make sure the recorder had enough space for the interview. "Everyone always wonders how people get up the nerve to do comedy. Not to mention that you know half the people in this town."

"Yeah, but most of them are sporting bad hair they blame me for." Judy ran her fingers through her own hair as she spoke. She felt the razor stubble at her neckline. "Thank God, I've got a good stylist." They both laughed.

"So," Lynn asked, "Can I turn this on?"

"Sure. I don't have spinach in my teeth, do I?" Judy checked her teeth in the back of her spoon.

Lynn laughed as she pressed Record and held the machine to her mouth. "Interview with Judy Parker, stand-up comic." Judy smiled. *That's me. Stand-up comic. Not high school drop-out. Not mediocre hair stylist. Not Mom, you're embarrassing me. Stand-up comic.* She liked the sound of it.

Lynn began the interview. "So, Judy, why'd you decide to try your hand at stand-up comedy?"

Judy put her fingertips together and rested her chin on top. "I've always done stand-up comedy. It's just that this time I got up on a stage. And made money!"

Lynn leaned toward the machine. "Judy won The Jest Fer Laffs comedy contest last Saturday night and will be among the finalists vying for Funniest Person in Texas." She sat back up, pushing her hair behind her ears. "Is it as scary as it looks? I get ulcers just thinking about it."

"Well, it was a little scary," Judy admitted as she twirled a fork between her fingers. "It's like the first time you have sex with someone new. You sweat a lot, you're afraid you'll do something wrong, and you wonder if they'll respect you in the morning." She stopped. "Oops! You probably can't use that, can you?"

“It’s fine. We’re the alternative paper. We can get away with things they can’t in that other paper,” Lynn said with a smile. She moved her chair a few inches to the left, out of the sun. “Where do you get your material?"

"Real life. You just can't beat it for funny. Most of the stuff in my set has either happened to me or friends of mine or clients at Lone Star Cuts."

Lynn interrupted. "How about that stuff about Dick Sloan? Was that real life?"

“You know, you were there,” Judy interjected.

Lynn nodded. “I’m playing reporter here. Just consider me an objective observer."

"Oh, sorry, Lois Lane!” Judy giggled. "I overheard his wife, who’s a regular client of the girl who works next to me, say he was impotent. I just thought it was funny for a guy who’s running for Governor, in Texas, of all places to have this specific problem."

"Especially considering his slogan?"

Judy tried to think if she knew what Dick's slogan was. "His slogan?"

Lynn couldn't hold back a laugh. “A Big Man for a Big Job.”

"Oh, that’s really funny. I wish you’d told me that before. I could have used it."

"Then you'd have had to share your check with me."

****

A few minutes later, Lynn turned the recorder off. She looked at her watch. "I think that'll do it. It’ll be in the Friday paper. You're gonna be a star."

Judy picked up her Dr Pepper, stuck the straw in her mouth and finished it off. "I could live with that!"

The Shit Hits the Fan

The day already seemed like a month, even though it was only 10:17, according to the clock slowly ticking the minutes away as if keeping time in a black hole on the other side of the galaxy.

Jimmy Joe stood with a cordless phone cradled against his right ear, the one with the hearing aid. He was shoeless, with one sock foot resting on a cheap brown plastic chair, the other on the floor. All ten toes were curled under as if he were a prehistoric ape trying desperately to hold on to a tree branch to avoid being eaten by the lions below. And that’s exactly how he felt.

He glanced over at the door to the alpha male ape’s office, then rubbed his neck compulsively. “““Believe me, you don’t want Dick gittin’ wind of this today!” he said hoarsely into the receiver. As he was hanging up, Dick slammed open his door and swaggered across the floor, looking like he was ready to fling poo at anyone in the vicinity. He clutched a copy of the *Travis Sentinel*.

Damn! Jimmy Joe thought. *Here it comes*.

"Have you seen this piece of shit?" Dick yelled as he slammed down the newspaper on the desk in front of Jimmy Joe.

"Uh-huh. I tried to talk to you about it when you got in."

"Who the hell does she think she is?” Dick interrupted, his blood pressure already so high his face was maroon, Texas A&M's school color. This is one of the things that always pissed him off. If he had to turn a color when he got mad, why couldn't he turn burnt orange? Damned Aggies!

"Calm down," Jimmy Joe begged. "You're gonna stroke out right here! Then a little impotence won't matter at all."

Dick shot him a look that would have seared the hide off an armadillo. Jimmy Joe just averted his eyes and slowly began to loosen Dick's grip on the paper. He spread it out on the desk and reread the piece.

While on stage, Ms. Parker took a few pot shots at gubernatorial candidate Dick Sloan. "I heard one of our clients say her husband is impotent, which is kind of funny, considering her husband's name is Dick. Dick Sloan. You may have heard of him. He's running for Governor... I think we should have a Governor who has to rely on the brain in his head, rather than the one in his pants."

Jimmy Joe had to stifle a laugh. Dick slumped into his chair.

"Well?" Dick demanded. "What are we gonna do?"

Jimmy Joe stood with his hands flat on the paper thinking. Thinking, *Wouldn't it be nice if I were lying on the beach in Padre Island? A tall, frosty pina colada in one hand, watching the waves rolling in and out.*

"Damn it, Jimmy Joe!" Dick said impatiently. "Call *The Sentinel* and issue a statement denying the charges."

A grin broke out on Jimmy Joe's face, but only because Dick was still staring out the window and couldn't see it. "It's the *Sentine*l, Dick. No one reads the *Sentinel*. And they're not charges. It was a comedy routine, for God's sake!" He attempted to unbutton his collar, but the button came off in his hands. Jimmy Joe stuck it in his pocket.

"I don't give a rat's ass what they call it!" Dick shook his fists like an angry baboon ready to pick a fight over territory. "I want a retraction."

"There's one tiny problem with that. The facts are right. I've seen you popping those little pecker picker uppers." Jimmy Joe looked Dick in the eye. "If the media looks into it, they can confirm it. Shit, just one call to your doctor." Jimmy Joe mustered on. "And if this comedian said she heard the story from Sarah, I'm sure your wife would probably be happy to just go on and on about your little problem, especially after that incident with the Rangerette. I told you to keep your hands to yourself."

Their conversation was momentarily interrupted by the sound of a siren. A fire engine went speeding by. *Stop here*, Jimmy Joe pleaded silently. *We've got enough fires to keep you busy for weeks.* And he wasn't just thinking of his hemorrhoids.

Dick stared out the window, instinctively attracted to other people's trouble. "We'll say they're for my blood pressure." He turned back around, picked up the section of the newspaper that was threatening his manhood, and rolled into a giant spit ball. His eyes darted around the room as if searching for someone to hit in the back of the head with it. "What's truth got to do with anything? This is politics!"

Jimmy Joe sighed. Dishonesty was something he'd had to learn on the job, unlike Dick, for whom it was as easy as ripping a carcass apart with his bare ape-hands. "We're gonna have to handle Sarah, too. We can't go denying it and have Sarah out there spreading the truth."

Dick un-crumpled the newspaper article. "Judy. Judy Parker. That's the bitch's name." He balled it back up. "You talk to my wife!"

"I think it'd be better if you talk to Sarah," Jimmy Joe said haltingly.

"Hell, we haven't spoken in weeks. I figure the less I say, the less chance the bitch will divorce me before the election in November."

"Assuming we make it to November," Jimmy Joe muttered to himself.

"How bad do you think it is?" For once Dick sounded almost pitiful.

"Well, I think a lot of people, women mostly, will think it's funny, especially the kind of women who read the *Sentinel.* You know the kind, hair under their arms, pierced tongues..." Jimmy Joe paused. Wild, unruly women were his type. "Most of those women wouldn't vote for you if you were running against a rattle snake. The question is, if those oil company execs and boot-scooting boys down on the farm get a wind of this, will they think you're not quite the man you say you are?"

Dick kicked the closest chair. It rolled three feet and slammed into a nearby desk. Almost to himself, he said, "We need a way to show off my manhood."

Jimmy Joe couldn't help it, he had to laugh. "What 'cha gonna do, pop a few Viagra, call a press conference and flash a bunch of reporters?"

"Shit, JJ, I'm serious here," Dick replied, looking as deep in thought as a man personally trying to disprove Darwin's theory of evolution can look.

Jimmy Joe drew in a sharp breath and started counting backwards from ten. He looked at his watch. "I'll think of something. Meanwhile, you're late to the Young Hispanic Republicans meeting. For God's sake, Dick, don't call 'em wetbacks or any of your other cute racist nicknames."

Dick wasn't listening. He had other things on his mind. "See what you can dig up on this Judy Parker person. And the reporter. Blow them out of the water."

"The story will disappear faster if you just leave it alone. They're nobodies. The reporter works for a paper hardly anyone reads." Jimmy Joe tapped his pen against his nose. "I think it's better if we don't make them into somebodies."

"Hell, I'm not letting a couple of bitches question my manhood and throw my campaign into jeopardy weeks before the primary without some kind of payback. Just do it." He picked up a box of campaign buttons and stormed out of the room.

"Whatever you say." Jimmy Joe looked out the window, watching Dick speed off in his Lexus. *Alice was right. I should have run for the hills screamin' when Dick asked me to take this job.* He fingered a rogue button in his pocket like a worry bead.

Comedy Can Be Tragic

Donna sneezed. The appointment calendar was covered with eraser crumbs, and so was her lap. She swept a few aside, as she tried to keep from getting tangled in the phone cord Judy was swinging nervously back and forth.

"They did what?" Judy practically screamed into the receiver. Donna swiveled around in her chair. The color drained from Judy's face. Her blue-mascaraed eyes stood out against her ashen skin like a bunch of blue-haired old ladies on the beach at Padre Island. "I'll be right there."

"What happened?" Donna asked as Judy slammed the phone back into the receiver.

Judy was in tears. Blue tears. "That was the principal at Nolan's school. His teacher saw two strange men talking to him at recess. They got in a car and drove away when she came over to see what was going on." Judy grabbed her purse from the drawer next to Donna's left knee and headed toward the door.

"Is he all right?" Donna queried, concerned. She finally unwrapped the phone cord from around her shoulders.

"Yeah, you know, nothing phases Nolan." She took a deep breath. "When the teacher asked Nolan what the men wanted, he said they were asking him questions. Like 'Does your mom's boyfriend spend the night?'"

"They didn't try to get him to go anywhere with them, did they?" Just a month earlier, a fourth-grade boy had been kidnapped from the playground of a local school. Judy shook her head, tears streaming down her face.

"Well, thank the good lord and lima beans he's all right. That's the important thing," Donna said consolingly. "You don't suppose The Evil One is behind this?"

"I don't think the Evil One has two friends who would be willing to go around harassing people for him. Shit, when he moved out, he couldn't even find anyone to help him move." Judy retrieved her keys from her purse. "Besides, he doesn't know what to do with the boys when it's turn to have them for the weekend." Judy used her index finger to wipe some of the smeared mascara from her cheek. "I'll see you tomorrow. Can you cancel my three o'clock?"

"Sure. Good luck."

****

Lynn struggled to get through the front door, carrying two full bags of groceries by their corners. She had forgotten once again to take the nice big canvas bags she had bought to save the planet. The corner of the heaviest bag, the one with the Lactose-free milk and ice cream. ripped just as she got her key in the door and two cans of cat food rolled down the hall toward the elevator. She plopped the bags down on the floor and retrieved the stray cans.

"Coco?" she called as she finally got everything inside. "I've got food. Not that I want to wake you from your nap or anything." Coco was nowhere to be seen. *Probably sleeping on my new shoes at the back of the closet,* Lynn thought.

As she started putting away the frozen foods – pizza, waffles and Rocky Road ice cream – out of the corner of her eye, she saw the message light on her answering machine flash. She ripped open a bag of Fritos and walked over to the machine. With one finger, she pressed "Play."

"Lynn, this is Ed." Lynn almost sucked down a corn chip. "I'm sorry to tell you this over the phone…" Ed paused, bolstering himself to deliver the final blow. "I'm going to have to let you go."

Lynn collapsed onto an oversized white chair like a chicken that had just been de-boned.

"That Person of the Week story you did on the woman comic. Dick Sloan's people called. They've threatened to sue the paper. You know we can't afford to go to court. Hell, we can't even afford to…" He paused. Lynn could tell from his voice that he really was sorry. "My hands are tied."

There was another pause on the tape. Ed was searching for the right words. "One more thing. I don't know how many people you've told about, well, uh, being, uh, you know…." Ed had never been comfortable knowing anything about his staff's personal lives. Sexual orientation, marital status, favorite color… "Sloan's people know. Don't ask me how. Just thought you should know that they do. I'm sorry. Really, I wish this hadn't happened. Sorry. Bye."

Lynn dropped her head into her hands. Just then, like a prop from a David Copperfield show, Coco appeared on the back of the chair. She grabbed the cat and cried into his soft fur.

Half an hour later, a very damp Coco went back to doing whatever it is that cats do when they disappear for hours on end. Lynn got up, tossed the empty chip bag into the trash, and dialed Judy's number.

****

Judy was sitting on her couch, folding Spiderman underwear and keeping her eye on the boys as they watched TV on the floor. The phone rang.

"I'll get it," Stevie Ray yelled, jumping up and practically running over Judy as he rushed to the kitchen.

"Stop right there!" Judy said in her best "I'm Your Mother and I Make the Rules" voice. "It's after nine. I'll get it. You two start getting ready for bed. It's later than I thought." Stevie Ray shot her a dirty look, which she saw but chose to ignore. Nolan shuffled off toward the bathroom sleepily. She smiled watching him pad down the hallway.

Judy took three steps to the kitchen and picked up the phone on the fourth ring.

"Hello? Hi, Lynn. I was going to call you. You'll never believe what…" Judy paused to catch her breath and heard sobbing on the other end. "What's wrong?"

"I've been fired!" Lynn said through her tears.

"What?"

"Fired."

"They downsized you?" Judy asked incredulously.

"Fired. Dick Sloan called the publisher of the paper and said he'd sue if they didn't can my ass, so they caved in." Lynn blew her nose.

"This is because of the interview!" Judy dropped the two pairs of underwear she had dragged in from the living room. "This is all my fault! I am so sorry."

"It's not your fault," Lynn said. She paused long enough for Judy to think that maybe she did blame her a little. "The interview was my idea. And my editor approved the damn story! In fact, he was thrilled about getting to stick it to Sloan."

Judy was momentarily wordless. "I'm sorry," she repeated.

"What am I gonna do, Judy?" Lynn asked, almost inaudible through her tears.

"Why don't you come over? I'd come to your place, but I can't get someone to watch the boys this late."

"No, it's too late. I'll just sit here and cry some more on Coco's shoulder. If I can find him." She sniffed. "Anyway, it's not tonight I'm... What am I gonna do with the rest of my life?"

"Everything will be all right," Judy said, although not as convincingly as she hoped to sound. "Maybe you could talk to Annie."

"Annie's in Seattle for a conference. I left a message on her cell phone." Lynn said. She was kind of worried about where Annie could be this late and why she wasn't answering her cell phone. Lynn's brain was good at playing worst case scenario, even on a good day. And today was definitely not one of those.

Things started to fall into place for Judy. "Today at school, two men were asking Nolan a bunch of questions. Like who am I sleeping with? I wonder if they were..."

"Judy, what are you talking about?"

"Never mind." No point getting Lynn more upset. "Here's what you should do. Try to get some sleep. And tomorrow, come down to the salon and we'll talk. I think I have an open slot between 10:30 and 11:00. It'll be all right. I'll see you tomorrow."

Judy hung up. She finished folding the laundry, while chewing on her front lip. *Damn,* she thought, *the impotent asshole strikes again.*

Lynn held the phone for several seconds, until the computerized woman's voice came on and reminded her, "If you want to place a call, please dial the number." She hung up and looked around the room for Coco. She finally spotted him on the kitchen table, licking the ice cream that had melted through the bag. *That's just great! My whole life's a mess!*

Will I Be Bloated in My Next Life?

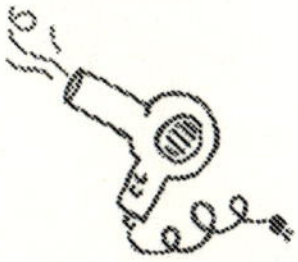

The Monday edition of *The Travis Sentinel* with the story reporting that Dick Sloan most certainly was not impotent – in fact, according to his medical records, he was as virile as he was at eighteen – lay coyly on Donna's desk, like it wasn't ruining lives all around it. Leon, Ima, Bernadette and Judy crowded around Donna's chair so closely that Donna felt like she was working in a phone booth full of college students trying to make the *Guinness Book of Records*.

Ima ran her French-manicured, amethyst crystal-adorned index finger across the headline, pausing occasionally to shake her head. Her long brass earrings tinkled. She looked up at Judy. "I warned you. Your moon is in retrograde AND you're bloated. That always means trouble."

Judy laughed half-heartedly. Ima wasn't joking, but Judy didn't believe in all that mumbo-jumbo new age stuff. Ima blamed the heavens for socks that got lost in the dryer.

Bernadette smiled sweetly but insincerely at Ima, then perused the article for herself. All trace of smile, faux or otherwise, disappeared. "Shit, what a vindictive dick! I say we get us a rope and fix his wagon." It was distinctly possible that Bernadette was serious. She had a red hot temper, not to mention a rope she kept in the back of her pick-up "for emergencies." No one had dared asked Bernadette what kind of emergencies.

"Yeah, let's lasso Big Dick by his Little Dick and haul him in for a bikini wax!" Judy said, only half-joking, thinking how much better she'd feel to see Bernadette ride off into the sunset with Dick Sloan dragging behind her Chevy half-ton. Bernadette was a big girl, she could probably get him hog-tied in no time.

Donna bristled and looked around the salon. "Judy, be careful!"

Judy shrugged, then looked defensive. "Hey, you all thought it was funny that night at Jest Fer Laffs! Don't bail on me now!"

"It was funny. Just not politically-correct. Hey, that's the first time I've really understood what that word means." Donna tried to roll her chair back a few inches to reach her box of tissues. She almost ran over Ima's sandaled foot with each toenail painted with a different sign of the Zodiac. Sagittarius and Scorpio were missing – not enough toes.

Leon leaned over Donna, his right hand unconsciously running across his left bicep. No one appreciated Leon's hard work at the gym as much as Leon himself. He read part of the story aloud:

Ms. Parker, a high school drop-out who cuts hair for a living, used 'comedic license' in her stand-up routine at Jest Fer Laffs. The story is not true, according to a spokesman for the Sloan campaign. In fact, the source said, Mrs. Sloan is not even a customer at Lone Star Cuts.'

"Yeah. I made it all up. Tomorrow I'm going to go around telling everyone I'm the Virgin Mary," Judy said sarcastically.

Leon snorted. "We can prove she's a client. They must know that."

Twenty feet away, Nan was busy cutting hair at her station. Judy glanced her way, then looked back at the group gathered at the reception desk. Oh yeah, Nan would make a real reliable witness. "*Yes, Judge, Sarah Sloan is a client of mine. Well, she's a client of one of my personalities. I think. We don't really talk all that much. What's that Helga? No, you can't have a Pop Tart!*"

"Here's the part that gets me. This whole mistress thing." Bernadette picked up the paper and waved it around as she talked. "According to Little Miss Anonymous, Dick is a stud-muffin. Says she's been sleeping with him for months."

"Maybe she means she's been sleeping near him. You know, snoring, dreaming, drooling, and laughing at his Gumby and Pokey PJs…" Judy trailed off.

"I wonder what poor Sarah thought when she read that?" Donna said, a hint of sorrow in her voice.

"I wouldn't worry too much about Sarah Sloan," Judy argued. "She's got looks, brains, money, and apparently, Dick's balls in her Gucci bag, so she'll be just fine."

Bernadette folded the paper in half and slapped it back down on the desk. "If y'all decide you want to string him up, come get me. I'm going out for a smoke."

Just then a pleading voice called from the back, "Leon, I must be dry by now. Come see if I'm ready."

Leon looped his thumbs through his belt loops and smiled that macho cowboy smile of his. "She's ready for me. They're always ready for me!"

"Leon, she's in her eighties," Donna said in mock horror.

"Yeah, Leon, give the woman a break. Her pacemaker might just give out." Judy added. Leon grinned and walked away. Ima followed, floating back to her manicure station.

Judy sighed. "Well, that's that. I'm never doing comedy again. It's too dangerous."

Ima stopped halfway back to her station and turned to Judy. "Good idea. Your horoscope also says you should avoid lawyers. And seafood."

"Does it say anything about lawyers bearing seafood?" Judy asked. Donna laughed. At least nothing had dampened Judy's natural wit. "Yep, that's the end of my comedy career. I don't want the papers next time to read, 'Judy Parker, aspiring comic, died on stage. Literally.'"

"You can't quit now," Donna objected. "You just need to try something a little less hot, politically. Besides, you won the contest. Aren't you supposed to go to the finals in Houston?"

Judy just nodded her head, then shook it sadly. All her dreams, up in smoke. Well at least she'd gotten $500 out of them. She glanced at the clock over the front door

"Where do you suppose your 9:30 is?" Donna asked.

"Oh, I forgot to tell you. She called when you were in the bathroom and canceled." Judy plopped down with on the couch next to her station with the grace of a beached manatee. Despite her remarkable weight loss, Judy still sat like a fat woman.

She sat and watched Nan work. Nan was exceptionally quiet today. It must be the personality Donna had nicknamed "The Librarian." Fortunately, it was the other personality who knew how to do hair. Nana's client, a thirty-something man with just the hint of gray in his beard, sat stone-faced as if being prepped for display in an open casket. Judy recognized him. He taught physics at U.T. and came in about every three months just to trim things up enough so he didn't look like a homeless mental patient.

Donna glanced at the schedule. "That's four cancellations so far this morning, all for this week "

Ima closed her eyes. "Let me see… Mrs. Strong, Mrs. Walker, Mrs. Hardin…?"

"...and Jolene Roberts," Donna finished. "You're really on today, Ima, psychic-wise."

"Fat lot of good it does. No one around here ever listens to me," she said looking at Judy who was busy peeling the red nail polish off her thumb. This sent shudders up Ima's spine just as if someone had walked upon one of her seventeen graves.

Judy looked up from her un-manicure. "You know what I'm thinking, Ima?"

"Of course I do. That all those clients are water signs and ..."

"No," Judy interrupted. "That they're all politicians' wives. And, all Republican politician's wives. I bet Dick Sloan is sticking his... Damn."

Donna finished erasing Jolene Roberts' name from the book. "They'll be back. Where else in town are these women gonna go for their Goodyear Blimp do's?"

Judy smiled half-heartedly. "I hope you're right."

The door opened and the chimes stared up again. "Hey, a customer," Judy said before glancing at who it was. Lynn.

Lynn squeezed onto the couch between Judy and Ima. "Well, this is quite a mess." Exhaustion oozed from her pores.

Judy patted her hand and looked at her. "You betcha. The bad news is, all my clients for today have canceled. The good news is, according to our calendar, Nan should be PMSing tomorrow, so I'll probably get all hers."

"Hey, Lynn," Donna said. "Sorry about what happened. The jerks!"

"Thanks."

"Did you see Bernadette out in the parking lot?" Judy asked. "She offered to string up Dick."

"Yeah, that'd feel good. Until we ended up in prison."

"Yeah, you need to stay out of prison. You're far too pretty. Although we could do something about that hair. While you're here." Judy replied.

"My hair is the least of my problems," Lynn muttered, pulling out the article from pocket of her jacket.

The reporter who conducted the interview, Lynn Larson, 37, a lesbian, is no longer an employee of The Travis Sentinel.

Lynn put the clipping on her thigh and smoothed it out with her palm as if this would make the words go away. "'Lynn Larson, 37, a lesbian.' Not exactly the way I wanted to be outed to my friends and family."

"How do you suppose they found out?" Judy asked. "You know it wasn't me. Really, I've been quiet as a damned church mouse or whatever religious rodent is supposed to be so quiet. It's been so long since I've actually been in church."

"I know it wasn't you," Lynn said, picking up the article and ripping it into tiny pieces. "I bet they got onto my computer at work and found out that I spend time on Gaydate." She began tossing the pieces over her head, like confetti.

"So, you think Dick Sloan can work a computer, huh?" Judy asked. "Not with his hardware problems."

Donna laughed. She couldn't help herself. Judy continued to peel the nail polish off her thumb. "Do you suppose it's over now? Or do they have other stuff up their sleeves?"

"Well, let's see. They interrogated your kids…"

Judy stared at her friend.

"Yeah, I figured that out." Lynn paused. "They got me fired. Humiliated us both in the paper. And they're apparently convincing your clients to cancel their appointments. What else could they do?" Lynn got up and went over to the mirror at Judy's empty station. She examined her face in the mirror.

The phone rang. Donna picked it up.

Judy watched Lynn. "Well, I dunno, they could make us wear thong underwear?"

Lynn sat down in Judy's styling chair and spun herself around. "Or they could put cement blocks on our feet and throw us in the Pedernales. I wouldn't put it past them."

"Yeah, big ol' cowboy boots full of cement. That's a look I don't know if even you could pull off."

"Judy, phone's for you," Donna interrupted.

Judy got up from the couch slowly. "Oh, great! Get that fan ready. There could be more shit ready to hit it." She walked over to Donna's desk and took the phone. "Hello?" she said tentatively. After hearing the voice on the other end, Judy smiled, put her hand over the receiver and mouthed, "It's Gail."

"Hi, Gail. Yeah, unbelievable, huh? I'll tell her you said so. Okay, sure. I'll call you. When this blows over. Bye." Judy handed the phone back to Donna and walked over to Lynn. She put her hand on her shoulder. "Gail said to tell you how sorry she is that you lost your job."

Lynn was busy staring at her reflection. "Maybe I would look good with shorter hair."

The phone rang again.

"Judy, you again."

"Great. When it rains, it pours." Judy took the phone from Donna.

"This is Judy."

"Hi Judy, it's Mark."

Judy's bad mood softened immediately. "Hi. Thanks for the flowers. I bet I'm to only girl in town to get a "Congratulations on your success/sorry you're being harassed by the next Governor" bouquet.

"Judy, my firm is doing some polling for the Sloan campaign."

"Shit!"

"My boss suggested I stop seeing you for a while, to avoid the look of impropriety…"

"Oh. I see. Well, I understand." Judy felt like the air had been knocked out of her.

"Judy, let me finish. I told him to mind his own fucking business!" Mark sounded proud of himself.

"You used the 'f' word? I've never heard you swear even once." Judy joked to cover her relief that she wasn't being dumped. At least not today and at least not on the phone.

"Well, sometimes a man's gotta do what a man's gotta do."

"What happened?"

"Oddly enough, nothing. He said to be careful, but left it at that. So, wanna hit the town tonight?"

Judy smiled. Dick Sloan may be after her, but she had friends and co-workers and now a boyfriend willing to take her side no matter what. *So go ahead, Dick, bring it on!*

Life Goes to the Dogs

Gail sat on the sofa listening to The Eagles' Already Gone, with Hans and Bubba, unwilling hostages on her lap. They'd much rather have been in the kitchen hounding for pasta, but she held them tight. She sang along with the CD.

"And so often times it happens that we live our lives in chains, and we never even know we have the key," Don Henley and the boys were wailing.

Everything around her was wet and salty. Gail's face, her butter yellow sweatshirt, the dogs' snouts. As the song reached the end, Gail released the hounds, who scurried into the kitchen to hit up Glen for handouts. Gail walked over to the stereo and turned off the CD player.

"Are you okay?" Glen asked as he walked into the living room, weaving his way around the dogs while carefully balancing a spoonful of spaghetti sauce in one hand. Gail slid the CD into the cover, looked up at him and smiled.

"Yeah. I will be." Gail pointed at the spoon and Glen walked over. She tasted and nodded. A little spicy, but hey, what was spaghetti sauce without a big dollop of salsa in it?

Glen looked at his wife and shook his head. "You know you don't have to do this. I really wish you'd just leave it alone," he said, staring into her big eyes. Hans and Bubba returned from the kitchen. Glen let them lick the spoon clean. The boys liked salsa.

Gail walked over to Glen and kissed him gently. "I have to." She patted him on the butt. "Maybe then I can finally get on with my life." Gail bent down and stroked Bubba's long ears. He licked her wrist. "It's like The Eagles say 'And so often times it happens that we live our lives in chains…'"

"'And we never even know we have the key,'" Glen chimed in. "I know, I know. Philosophy 101 by The Eagles. I could hear it in the kitchen. Hell, I'm pretty sure the neighbors could hear it down the block." Glen squatted down, Dachshund level, to get Gail's attention again. "But look what happened to, what's Judy's friend's name?"

"Lynn."

"Yeah, look what happened to Lynn. Do you want to lose your job too?"

"I don't think that'll happen." Suddenly the tears disappeared and a smile appeared. "If I do get fired, Judy and I will open a combination beauty salon/mammogram office. We'll call it Big Hair and Big Bazooms."

Glen laughed despite himself. He dropped the spoon on the coffee table, grabbed his wife's hand, pulled her up and held her close. She put her hands on his hips and sank slowly into his chest. "Sure, it's funny now. But if you piss Dick Sloan off, he's gonna have to put you in your place. Don't want the little wimmin folk gettin' out of hand."

Gail winked. "Oh, let me get out of hand just this once and then I'll take off my shoes and let you knock me up."

"Yeah, yeah. Make promises you don't plan to keep." He patted her thigh and sighed. For better of for worse, he reminded himself. Glen had a feeling for worse might be just around the corner.

Bad Boys, Bad Boys

Dick Sloan's voice could barely be heard over a bad instrumental version of 'Hey Jude' playing in the hall outside the ballroom of the Austin Hilton. Gail stood in the hall, next to a fake ficus, wearing a knock-them-dead gold beaded sheath dress. She smiled. Not at the B.S. Mr. Sloan was feeding the crowd inside, but at the idea of being here. Doing this. Getting it over with once and for all.

Glen, in a dark blue suit that Gail had picked out for him two years ago when he was best man at his brother's wedding, glanced at his wife nervously. Confrontation made him tense. And queasy. Not to mention that it made him have to pee. But from the applause and cheering coming from the ballroom now, he knew he didn't have time. To pee, that is. Tense and queasy he had time for. He flashed a supportive smile at Gail and said a quick prayer.

Suddenly the ballroom doors opened and several local reporters burst through. Having rehearsed his lines over and over in the car, Glen spoke loudly to the group of them. "Excuse me. My wife has some information about Dick Sloan you will be interested in hearing."

Two TV reporters and a writer for *Texas Monthly* stopped in their tracks. A still-employed reporter for *The Sentinel* continued down the hall.

Gail cleared her throat, "Dick Sloan date-raped me," she said unhesitatingly.

The Sentinel reporter who was now halfway down the hall turned on her heel and immediately hustled back to the fray.

"What did you say?" asked the Channel 11 reporter.

Gail took a deep breath. "Dick Sloan, the guy in there running on a platform of family values and moral principles? He date-raped me in high school."

The small group let out a synchronized gasp, as if they were the chorus in a long-forgotten Greek tragedy. The Channel 9 reporter motioned to his cameraman to turn the camera on. *The Sentinel* reporter clicked her tape recorder to Play and held it just inches from Gail's mouth. "This is a helluva charge. What exactly did he do to you and when?"

Glen reached out and grabbed Gail's hand, as much to settle his nerves as hers. Gail squeezed his fingers. "It was a long time ago, 1981. But that doesn't change anything. He attempted to force himself on me. I kept yelling 'No' and he kept yelling 'You ugly bitch!' I had to fight him off with my purse."

While Gail was talking to the spellbound group of news hounds, Dick and Sarah Sloan emerged through the ballroom doors and walked out into the hall. They were together, but not together. It's a skill married couples have who hate each other but are required to put on a show. When Jimmy Joe had begged Sarah to please tell reporters she hadn't said anything about Dick being impotent, she balked. She had only agreed not to answer the question. That was something. Not enough for Dick, but Jimmy Joe knew how gracious she was being to agree to that.

Dick stopped first when he reached the throng of reporters. Sarah pulled up ten feet behind him.

Gail paused for a second and looked directly at Dick. She hadn't seen him in person since high school, yet still his face haunted her nightmares. Even though he was fatter, balder and meaner-looking than in all those years ago, that face and those stubby fingers sent an involuntary shiver down her spine. She could feel her hackles raise. She glanced at Sarah. Gail felt sorry for her.

Not about to be stopped now, Gail once again returned her steady gaze to the reporters and continued with her story. It was now or never. "It was 1981. I went to the same high school as Dick Sloan and Sarah Davis, Mrs. Sloan. Travis High School. You can check out the yearbook."

Glen produced the 1981 Travis High School yearbook with the appropriate pages marked. Dick's smarmy face graced page 117. Gail's much younger, but very much the same, visage was on 119.

"What's this all about?" Dick hissed. For the first time, the reporters noticed him standing there. They encircled him like hyenas bearing down on their prey. The Channel 9 camera swung so quickly from Gail to Dick, the camerawoman would be nauseous for days. In fact, she worried for three weeks that she might be pregnant.

"This woman claims you date-raped her in high school," the Channel 9 reporter yelled to Dick, as if he were hard-of-hearing.

Dick's fists clenched and his jaws tightened. Sarah stood awestruck, like an armadillo caught in the headlights of a semi-truck on the freeway. An armadillo with an impeccable sense of style. Gail joined the circle of reporters, Glen still holding onto her hand. Her fingers were cold.

Dick sputtered, "That's ridiculous." He looked around the room until he spotted her. "I don't even know this woman!"

Gail raised her eyes to Dick's. "It's Gail Addison. Back in high school it was Gail Wyker."

"Do you have a response to her charges?" queried a *Statesman* reporter.

"Yeah, I have a goddamn response," Dick said angrily. Sarah backed several more feet away from him. She'd like to have left all together, but forty years of southern etiquette, added to a perverse desire to see how this played out, glued her feet to the carpet. Dick went on, "Obviously, my opponent has staged this whole thing. It's nothing but vicious mudslinging. Just proves the type of person Rich Clemson is."

Gail said quietly, but firmly. "We went to a drive-in movie. The show was *Raiders of the Lost Ark*. You had a bet with your buddies. Is any of it coming back to you yet, Dick?"

It may not have shaken any cobwebs loose in Dick's mind since it was cluttered with the residue of all kinds of lies and political mistruths, but vague memories started coming back to Sarah. "Shit, Dick. You..." she sputtered as the camera swung her way.

Dick interrupted. "Hell, maybe I did date her. I dated a lot of girls in high school. What can I say, I was a popular guy. But the rest is bullshit. Someone call security."

The Channel 11 reporter turned back to Gail. "Can you back up your story? "

"Yeah, it was, what, twenty-six years ago," *The Statesman* reporter added.

"That changes nothing." Gail whispered emphatically. There was a long silence. No one could hear Dick's blood pressure rising, Glen's stomach churning, or Sarah's heart finally closing the door on this chapter of her life.

"Well, I can tell you that while Dick was trying to rip my clothes off with his stubby little fingers, he whipped out his, well, let's say, alter ego and…" she hesitated for a brief second, "Attempted to have intercourse with me against my will."

"Bullshit!" Dick spat out vehemently. "She has no proof."

"You want proof? Let's say that I know that Dick has a birthmark about yay big," Gail held her fingers about an inch apart, "Shaped a little like a star, at least that's what he thought. It was on his thigh right next to his balls." Gail had never been able to wipe the image from her mind.

Sarah gasped loudly and the camera swung her way.

Gail, determined to finish what she started, plugged away. "He was mighty proud of that star. Said it meant he was going to be a Dallas Cowboy."

Sarah played nervously with her wedding band. "I remember you. You had really short hair and thick glasses. You were in the math club? You were a brain."

Gail laughed. "Not brainy enough to know better than to fall for it when Dick supposedly wanted to date me. I found out later it was all a big joke with his buddies."

It all came back to Sarah like a slow-motion nightmare. Not knowing whether to confront her spouse or console the poor woman, she left. It wasn't a choice as much as the Fight or Flight hormones kicking in. And she just didn't have any more fight in her. For now. She pushed her way through the semicircle of reporters. Before she entered the stairwell, she turned and gave Dick a look that said, "It's over!" Actually, that's not what it said at all. What the look said was, "You scum-sucking bastard. I hope they boil you in oil and bury you alive out in Big Bend where fire ants sting you almost to death and then the turkey vultures slowly peck at your filthy hide." No one followed her down the stairs.

The cameras swung back to Dick again. After several seconds of deafening silence, he spat out "I have no comment."

Gail smiled. Glen finally took a breath.

"Stop tape. Let's get back to the station and edit this. Wow, Harry's gonna love it." The reporters rushed out. Dick stormed back into the ballroom and began dialing his cell phone furiously.

Still standing together in the hallway, hand-in-hand, Glen looked at his wife worriedly, "Are you okay, hon?"

"Actually, I feel great. I feel like celebrating," she answered with a smile.

"Celebrating?"

"Let's go dancing." Gail did a pirouette.

Glen took her hand and twirled her around. "Dancing, huh? Well, okay. It'll be the perfect end to a perfect evening – appear on TV, ruin a political career, and go dancing."

Who Says There's Nothing Good On TV?

Lynn slouched on her sofa, eating ice cream from a gallon container with a scoop. Coco slept quietly at the other end of the couch, his tail flicking occasionally, swatting off dream dragonflies.

Listlessly, Lynn raised the ice cream carton to her forehead and held it there for a few seconds, hoping the icy coldness would sooth her raging headache. She got migraines every time she ovulated. Damned eggs.

When she removed the carton from her forehead, her attention was drawn to the TV screen across the room, where Dick Sloan's angry face appeared, looking like the poster boy for road rage. *Asshole!* She thought as the pain in her head stepped up another notch. But, it occurred to her, if he's angry, maybe it was good news. She turned up the volume.

"In a late-breaking development in the Sloan campaign this evening, a woman claimed that Mr. Sloan sexually assaulted her in high school." Suddenly, Dick was no longer on screen, but Gail, Judy's new friend was.

"Oh, my God!" Lynn muttered as she sat up straight and leaned in toward the set to be closer to the action.

Gail's voice was saying, "He attempted to force himself on me and when I said 'No!', he continued, all the while yelling, 'You bitch!' I had to fight him off with my purse."

"I can't believe this!" Lynn looked over at her dozing cat. "Can you believe this Coco? I know this woman!" Coco awoke but was not impressed with the news.

The phone rang. For a second Lynn thought it was on TV, but she quickly picked it up. "Judy. Yeah, I'm watching now. Wow! She is one ballsy woman. Yeah, I know you did her hair. Sheesh. Quiet."

The TV reporter's voice came on again. "The woman, Gail Addison, formerly Wyker, a radiological technician at Mercy Hospital, claims that she can prove her allegations. She claims Mr. Sloan has an unusual shaped birthmark located in a private area. Mr. Sloan denies all charges, calling them 'Damned lies.' Mrs. Sloan, who witnessed the startling confession, left angrily after Ms. Addison made her charges."

"I'll bet she did!" Lynn said to the television, momentarily forgetting that Judy was on the phone. "What? Sure, you call her. Then call me right back."

Lynn bounded up off the couch towards the freezer with the ice cream. Her headache had miraculously disappeared. Coco looked annoyed at all the commotion. A cat's gotta have twenty hours of beauty rest or the world's gonna have hell to pay.

Join the Uppity Women's Relocation Program Today!

Naturally blonde hair piled up at the corner of Judy's station as her broom glided across the floor noiselessly. Just a few feet away, Nan, wearing a firefighter's hat two sizes too big, was busy trimming a client's hair, also blonde, although there was nothing natural about it. Unfortunately the hat kept slipping down over Nan's eyes making it hard to see what she was doing and causing her to occasionally nip her client's ear.

Ima was busy doing nails in the back. Leon had one client under the dryer and was admiring himself in the mirror above the shampoo basin as he rinsed another client's soapy hair. Bernadette was in the middle station, perming the hair of a forty-eight-year old legislative aide, male, well-dressed, Republican, and gay.

It was a busy day at Lone Star Cuts and about to get busier.

The chimes rang their incessant welcome and Gail walked through the front door. She took a look around, then caught Judy's eye and said, "Whoa. Looks like business has picked up!"

Donna, having just reappeared from the bathroom, tissue in hand, waved. "Hey, it's our hero!"

Judy walked over to Gail and hugged her. "You're still alive! I'd have thunk Dick would have run you down in the parking lot of the Hilton last night."

"Nope, not a scratch on me. Apparently he's much better behaved when there are cameras around," Gail said.

"Then I'll just have to get Nolan a video camera to mount on his bicycle." Judy smacked her gum. She joked, but the fear cut to her heart. "So, what are you doing here? Shouldn't you be at work feeling women up? Or did Dick get you fired too?"

Gail shook her head. Her dog earrings glinted in the sun. "Hell no! I took the day off. You wouldn't believe how much better I feel getting that awful secret off my chest." Gail did look lighter and more buoyant.

To quell some of her giddy energy, Gail sat on the sofa nearest Judy's station. Judy stood leaning against her chair. "I gotta give it to you, you do know how to get attention. I am so happy we're both on the same side, you bold woman!" Judy looked at her brave new friend and continued. "And can I say, you looked stunning on TV last night. That hair. You know I'm taking full credit."

They both laughed. Donna too.

Judy paused, looking for any signs of fear or regret. "Are you worried about what might happen now?"

Gail smiled. "Hey, Dick can't hurt me any more now than he has for the past twenty-six years. For the first time since high school I feel like I've gotten rid of him." Gail looked around the salon. She noticed Nan's fire hat and motioned to Judy questioningly. Judy shrugged and whispered loudly "Either Fred the mild-mannered obsessive-compulsive has taken up fire fighting, or Nan has just given birth to a new personality."

"Gee, I wish you'd told me, I'd have brought a gift," Gail joked. Judy smiled. Gail crossed her long, freckled legs and smoothed her ankle-length royal blue skirt that would have hung to the floor on Judy. "Anyway, if Dick butts into my life, you and me are going into business. Big Hair and Big Bazooms. Although I guess we'll have to join the Uppity Women's Relocation Program first and change our identities."

"I think I have the number in my Rolodex," Judy said with a giggle. "Big Hair and Big Bazooms, huh? I like it." Never one to be outdone by an amateur, Judy came up with a business idea of her own. "Or, we could start an auto parts store just for women," Judy suggested as she dumped the hair into the trash can. "In the mall. Next to Victoria's Secret. We could call it Le Auto Parts Boutique. All the products would be color-coordinated. Mauve for Honda Accord, Pearlescent Pink for Cadillac, Burnt Sienna for Jeep 4x4..." She paused, thinking. "And all the fluids would have different scents. That way, we'd be able to tell just by sniffing whether we're leaking Calvin Klein Antifreeze or Oscar De La Renta Transmission fluid."

Gail broke out into laughter so hard her breasts shook like large bowls of Jell-O duct-taped under her bra. "God, Judy, way to take my mind off things! That's why I came here. Just in case reality started to settle in and smack me across the head and say, 'What the hell were you thinking, girl?'"

Judy, remembering Gail's coffee mug, said, "Hey, why let reality ruin your day?"

"Exactly!"

"One more thing," Judy added. "We all know that given a choice between 30-weight and 40-weight, women will choose..."

Gail and Donna responded together, "30-weight!"

"Right!" Judy grinned. The audience was with her. "So oil will be sold in petite, medium and full-figured sizes."

"Great stuff. Is that for the Houston comedy finals? " Gail asked.

"No!" Judy's smile disappeared. It was replaced by obvious disappointment. "No more comedy for me."

Gail stopped laughing. Her eyes locked on Judy's and held them. "Oh, you'll be up there again. It's something you have to do. Take it from someone who knows. Some things you just can't control."

Judy smiled again, unsure. "You're much braver than me. I could never do what you did last night."

"Funny, I thought you were the brave one. You'd never get me up on a stage in front of a bunch of people and try to make them laugh," Gail shook her head in wonder.

Judy laughed. "Whaddya mean? There must have been several million people watching you last night!"

"Yeah, but they weren't there where I could see them." Gail paused. "And I wasn't trying to make them laugh."

"I don't think anyone was laughing. Least of all Dick," Judy agreed.

Donna added, "Or Sarah. I don't think Sarah was laughing either."

Gail looked at the clock. "I am sorry she got caught in this mess."

"Hell, she can't stand him anyway," Judy interrupted. "You did her a huge favor. I wouldn't be surprised to see the divorce headlines tomorrow."

"Well, then I've done my job," Gail said as she stood up. "I really should go. I have a massage appointment and then I'm going to feed the ducks at Zilker Park."

"That's right. Keep moving. A moving target is harder to hit." Judy picked up her scissors and snipped them in the air. "After that job you did on Dick last night, maybe I won't need to use these on him after all!"

Gail was just about out the door when Judy called to her. "Hey Gail, I have an idea. Lynn needs consoling and

I'm thinking maybe we should throw her a party. You and me, the survivors of Dick Sloan." Judy paused. "Of course we'd have to have it at your house, because mine's not fit for humans. At least not any over the age of fifteen."

Sex, Lies and Spin Doctors

Something smelled and it wasn't the stale cheeseburgers and fries that were sitting on the volunteers' desks because they'd been too busy fending off angry constituents to snag a few bites. The smell was fear. Had there been a dog in the room, he'd have put his nose in the air and then rolled in it, trying to cover it up immediately

Jimmy Joe smelled it. In fact, he was producing a lot of it. But not nearly as much as Dick. Dick was a one-man fear manufacturing plant. He was rolling fear out like he was months behind in filling orders. "Where the hell are those thirty-nine cases of Shit Scared Out of Me I ordered on the thirteenth?" an angry customer yelled from Dick's sweat glands.

In the outer office, Jimmy Joe paced frenetically from desk to desk, like a caged panther. Every time the phones rang, which they had been doing every few minutes since he opened the office at six a.m., Jimmy Joe would practically jump out of his skin. No easy feat since Jimmy Joe's skin was the texture of coarse sandpaper.

Jimmy Joe stopped his pacing momentarily, and stood behind one of the newer volunteers, a retired engineer, to listen in on the one-sided phone conversation. "The reports are completely false. The woman's not mentally stable. She's been in therapy for years. You can't trust anything she says…" Jimmy Joe had scripted the response at two a.m., three hours after the shit hit the fan at the Austin Hilton. It was important that they all be on the same page. He was pretty sure they were going down, but he wanted to make sure they at least had their story straight. The volunteer hung up. "Good," Jimmy Joe grunted, trying to sound optimistic.

Suddenly, Dick yelled from his office. "Get in here, Jimmy Joe. Now!" The glass in his door shook. Jimmy Joe dragged himself slowly and insolently toward Dick's office. It was all he could do not to run screaming out into the street. Instead, he took shallow breaths, opened the door, and looked Dick square in the eye.

"Well? What the fuck are we going to do now?" Dick demanded, trying to cover the smell of fear with Stupid Macho Man, a fragrance that was equally unappealing. Jimmy Joe wouldn't have been surprised if, at that very moment, Dick had pulled out a gun and told him to go off the redhead and her husband.

Trying to get his blood pressure down, Jimmy Joe took a deep breath before answering. His exhalation sounded like the air being let out of a very large balloon. It came rushing out and kind of squeaked at the end. "We're doing all we can. It helps that she has been in therapy. We're spinnin' that for all it's worth." He paused to breathe again. It didn't seem automatic this morning. He kept having to remind himself to do it. "But that doesn't change the fact that we're three days away from the primary and not only do we have more bad PR, we're also almost broke. You better find your wife and make nice. Again."

Dick grunted. He looked exhausted and afraid for once. And briefly, ever so briefly, Jimmy Joe felt sorry for him. But it passed quickly.

"You find Sarah," Jimmy Joe said as he started to close the door behind him. "I'll get all the volunteers to pray real hard. It couldn't hurt."

The Survivors of Dick Sloan

Gail scurried around the living room, trying to make everything perfect for her guests. She was a natural born hostess, but this was the first time she'd thrown a party like this – a half celebration/half condolence party. She had once thrown a baby shower for one of the nurses at work who'd found out she was pregnant with twins by her ex-husband after he started hormone treatments to become a woman. That had kind of the same feel. But this time there wouldn't be any baby stuff to ooh and ah over. There was, however, a giant picture of Dick Sloan that Gail had turned into a dart board. And she did know a woman who worked in State Senator Juarez's office who mentioned that they were looking for a new communications director. So there was possible good news for Lynn.

As she looked around the room, Gail smiled. She was the kind of woman who had had lots of friends in her life. But less than a year ago, her best friend Beth had suddenly died of a heart attack at only forty-nine. Between Beth's multiple sclerosis, the medical bills that were piling up, and her adopted autistic eight year-old son, Dermott, who had tried to kill her twice in her sleep, the heart attack wasn't all that unexpected.

The loss of Beth had left a hole in Gail's heart. And since her heart was big, the hole was big as well. So the chance to make new friends, especially one like Judy, who made her laugh as much as Beth had, was something Gail couldn't be happier about.

The cocktail wieners speared with red toothpicks were lying on a silver platter in the middle of the coffee table. Subtlety would not be the theme tonight. There were Margaritas in the blender and Glen's semi-famous homemade southwestern lasagna baking in the oven.

Gail moved one of Bubba's toys, a now squeakerless stuffed squirrel, from next to the sofa. The doorbell rang. She hurried to answer it.

Judy and Lynn stood on the other side, all smiles. Gail was surprised at how pretty Lynn looked without a scowl on her face. Her features were softer and she looked much younger.

"Before I can let you in," Gail whispered, "You have to tell me the secret password."

Judy laughed. "Beans and weenies."

Gail opened the door wide and let the duo in.

"Wow," Judy said. "This is much nicer than my place. Although I do have that nice sunken living room."

"Really?" Gail asked.

"Actually, she has a hole in her floor. I've offered to get someone to fix it for her, but apparently Judy considers it a conversation piece."

"Not to mention how much easier it is to take out the trash." Judy winked at the two women. "Speaking of trash, love the dartboard, love the teeny weenies!"

Lynn shook her head. "I'll pass on the wieners, thanks."

Judy laughed. "Well, more for us then."

"What smells so good?" Lynn asked.

"That's Glen's special lasagna. He's a great cook. Thank God because if we had to rely on my culinary skills, we'd be eating peanut butter and jelly every night" Gail said smiling. "And the dogs aren't supposed to have peanut butter. It makes them fat."

"Stevie Ray and Nolan love peanut butter and jelly," Judy piped in. "Maybe you can come over and cook for them sometime."

"Sure, why not? It'd be nice to have someone who appreciated my special gourmet talents." Gail ushered her guests to the sofa. "Hans and Bubba were gracious enough to let us use the sofa tonight."

"Hans and Bubba?" Lynn asked.

"My dogs. They're out walking Glen. He said he'd stay away until I called him and told him it was safe. I think he was afraid we'd get on an "All Men are Evil Scum" tear and he'd suffer the consequences."

"Well, with this one here…" Judy started, but Lynn interrupted.

"Hey! I like some men. My brother. The guy who reads my electric meter, he's okay."

"That's it, just two men make the cut?" Gail asked. She liked men in general. Just not the skuzzy ones who gave the rest of them a bad name.

"Hey, I'm not here to drag this party down," Lynn said. "Dick Sloan got what was coming to him and eventually I'll find a new job, so let's party, ladies."

Gail got up and headed towards the kitchen. "Actually, Lynn, I may be able to help on that front. What would you think about working for a good politician?"

"Is there such a thing?" Lynn asked, following Gail, eager to hear her out.

Judy jumped off the sofa. "Hey, you two are not going in there and talk politics. We came here to get drunk and get to know each other better. Or vice-versa." She stopped for a second. "Although considering both of you have seen me at least half-naked, I don't know if you can know me any better."

Gail and Lynn returned to the living room, Gail carrying a small pitcher of green frosty liquid. She poured drinks from the blender into the blue-rimmed Margarita glasses she'd set out on the coffee table. She handed the first drink to Judy. "Yes, it's important that we get to know each other, because if we do end up having to go on the lam for our lives. I don't want to find out then that you snore horribly or walk around naked in your sleep."

Lynn smiled as Gail handed her a drink. "Hey, I don't object to naked sleepwalking."

"See these are the things we'd need to know," Judy said. The trio laughed as they all settled into Hans and Bubba's sofa.

"How long 'til your man's lasagna is ready?" Judy asked hungrily.

"About thirty minutes."

"All right, then. Gail, tell us everything we need to know about you in ten minutes or less. Starting now."

"Oh, no pressure." She thought for a second. "Okay, I'm a petite blonde who enjoys bareback riding on the beach..."

Judy laughed. "Stop it! That's Lynn's fantasy woman you're describing. We don't want her getting all hot and bothered."

"No, no," Lynn laughed. "Keep going. How petite? How blonde? What kind of horse?"

"I see what you mean," Gail said taking a sip of Margarita. They were perfect – limey, but not too limey. She may not be good with a whisk or a frying pan, but Gail could wield a mean blender. "Okay, the real me. Let's see, married, two Dachshunds instead of kids. I had planned on becoming a rocket scientist, but decided to become a radiologist instead..."

Lynn interrupted. "What changed your mind?"

"You sound like you're doing an interview. Is there a tape recorder hidden in your purse somewhere?" Gail asked.

"Sorry. Force of habit." Lynn apologized. "But I am interested. It always fascinates me to know why people choose their careers. I wanted to become a journalist because I was one of those kids who started asking 'Why?' when I was three and never stopped. Judy decided to do hair because it's the only job where she can dress tacky and talk about herself."

Gail looked at Judy to see if this hurt her feelings. Judy just started laughing. "Yeah, that's pretty much it. It was either hair or my own talk show. When Oprah retires, look out world!"

Lynn steered the question back to Gail. "So, what was it that made you decide to become a radiologist?"

"Well, I was always the nerdy girl in high school and college. I knew I wanted to do something in science. I wasn't kidding about the rocket science part, but when I thought about the fact that most research in that field is about killing people and destroying things, I knew that was out. I like people. Men and women."

Lynn smiled. "Me too."

"Not in that way. Okay, maybe just a little. They are nicer to look at, generally speaking. And believe me, I know."

Judy laughed.

"Don't you think so?" Gail asked.

Judy nodded. "Yes, generally speaking. Except for that one part I enjoy a lot that we women are missing."

"I'm not talking about having sex," Gail said.

"Darn!" Lynn said, her eyes twinkling.

"Okay, I'll admit it. When I need something to help me get in the mood, I usually look at the Victoria's Secret catalog. I tried naked pictures of men, but they just don't do it for me. There, are you happy Lynn?" Judy asked.

"I knew it all along. Every woman has an inner lesbian."

Gail nodded. "You're probably right. I subscribe to *Playboy*. It's in Glen's name, but he just reads the articles usually, unless there's a really good-looking redhead. He has a thing for redheads." Judy and Lynn could sense the love in her voice. It was nice to see a woman so in love with her husband. Especially when there seemed to be so many who were not.

Judy laughed. "Well, that's convenient!" There was a pause in the conversation. Gail noticed they had all finished their drinks, so she sashayed back to the kitchen for the extra pitcher she'd stuck in the freezer before the girls had arrived.

"I like her," Lynn said.

"Well of course you do. She's cute, funny, and now you've got that whole lesbian fantasy thing working for you."

"Besides that. She's just so open and…"

Gail returned. "Were you talking about me?

Judy nodded. "It was all good. And Lynn promised not to put any moves on you. I told her Glen would sic the dogs on her if she did."

Gail refilled Lynn and Judy's glasses, then her own. "What were we talking about before we got off on sex?"

Judy couldn't suppress a giggle.

"Oh grow up, Judy! You're like a seventh grade boy." Lynn scolded her, but then started giggling herself. "I believe I had asked you why you became a radiologist."

"That's right," Gail said. "Well, when I gave up the rocket science notion, I wondered what I could do that was scientific and would serve some larger purpose. One of my professors in grad school got breast cancer…"

Judy interrupted her. "You went to grad school?"

Gail nodded. "I got a masters degree in biology before I went into radiology."

Judy elbowed Lynn. "Well, I guess you're not Miss Higher Education any more, are you?"

Lynn shook her head. "You know it doesn't matter to me that I went to college and you…" she paused for effect, "went to beauty school. Barely."

Judy raised her lips and hissed, extending her long red fingernails, freshly painted by Ima.

Gail shook her head. "I believe in higher education, but I don't think you have to go to school to get it. I bet Judy knows stuff most people with Ph.D.s will never understand."

"You got that right, sister," Judy said with pride. "The stories I could tell."

"And have," Lynn and Gail said at the same time.

Gail continued. "Besides, from what I've seen, most professors are really good at one thing, but they can't do anything important like buy groceries or fill their car with gas."

Judy suddenly got quiet. "What's the matter?" Gail asked.

"It's just, well you mentioned your professor got breast cancer. Made me think of my mom."

"I'm so sorry about your mom, Judy." Gail said sincerely. "I don't know any woman who hasn't been personally affected by breast cancer in one way or another." Lynn patted Judy's hand. Gail continued, "When I found out my professor was sick, well, that's when I chose radiology. I thought if I could help women find cancer early enough I could help save their lives."

"You are so noble," Lynn said.

"Not really. It's really all about me. I feel better about myself when I'm helping someone else. So it's all about me, me, me."

Judy shook her head. "Nope, sorry, you are noble." She turned to Lynn. "I don't know if we can be friends with her. She outclasses us in every way."

Lynn nodded in agreement, then paused. "Well, we do have that *Playboy* thing on her."

Judy laughed. "Yeah, that's something!"

Winning the Battle

Despite everything – the lies, the misdemeanors, the accusations, and the talk around the capitol – Dick Sloan won the Republican primary. After all, it was Texas, the state where Tom Delay convinced voters of his ethics for decades.

The primary, however, was too close for comfort. What should have been a walk in the park had ended up a bloody battle in the back alley, and Dick and his staff had spent almost all their campaign funds on brass knuckles and political goombahs.

Sarah Sloan didn't attend her husband's victory party. That fact was duly noted in all the major statewide media, including the newspaper that had once employed a certain uppity lesbian reporter. In fact, the day before the party, Sarah had yanked all her money out of his campaign, kicked him out of the house, and filed for divorce. Then she got her hair highlighted by Nan. In fact, she'd called the media anonymously, leaving a tip that Sara Sloan had an appointment at First Cuts, which was her usual beauty salon. She felt younger, happier, and lighter than she had in years. And the part she thought would scare her, the *being alone without a man at her age* part that was made much easier by the fact that she had lots of money and plenty of friends who had hated her soon-to-be ex, and were more than happy to sip champagne with her and discuss his shortcomings night after night.

Lynn heard the news of Dick's victory second-hand from Judy, because at the time of the announcement, she was in Denver meeting Annie for the first time. Annie, who was petite and blonde and loved to ride horses. Lynn hadn't yet checked under her bed to see if she subscribed to *Playboy.* But it didn't matter, because Lynn was definitely in love. It had to be love because her head spun, her stomach churned, and just the sound of Annie calling her name made her warm all over. So, though she was incensed that the bastard had won the primary despite what he'd done to Gail and Judy and her, she still couldn't keep from smiling. It was a stupid, lopsided grin that didn't look like it belonged on Lynn's perfect face.

Gail had seen Dick's smirking face on late night news and sighed when the anchor woman announced the final results. But she chalked it up to a part of her life that was over now. So she patted her sleeping husband on the head, turned off the TV, moved the dogs to the end of the bed, and went to sleep.

Mark was at Judy's house when the news broke. But he'd already told her it was going to happen – he'd seen the polling numbers. It seemed the good ol' boy network wasn't bothered by the date rape charges. In fact, they may have balanced out the impotence story.

Judy turned off the television and stared into Mark's eyes. She was lucky she had him. And she was starting to really fall for the guy. She kissed him hard on the mouth, grabbed his hand, and tugged him towards her bedroom. The boys were at the Evil One's so they could make all the noise they wanted for a change.

"Now? I thought you'd be upset!" Mark asked.

"I don't care about the jerk. At least not tonight." Judy ran her fingers over Mark's broader, sturdier ones. To hell with bad boys. Let them become Governor. She was sticking with the good guys from here on out.

As they made love that night, Judy had four orgasms, a record. And when she pulled her fur-lined handcuffs out of the nightstand and Mark smiled at her, she almost had another.

Close Shaves and Lap Dances

"Anyone ever tell you that you look just like George Strait?" Nan whispered huskily in her client's ear. The client, a loan officer at Austin City Savings, was either oblivious to her advances or was doing a really good job of ignoring her. Which was hard to do, especially today, since she was wearing red flannel Garfield pajamas and black cowboy boots. Pretty unusual even for Nan, considering it was at least 90 degrees out and Nan tended to get heat rash. Although not all of her personalities were susceptible.

Nan shook a bottle of mousse and deposited a large dollop into her right hand. Then she straddled Roger's knees as if getting ready to perform a lap dance for him. "So, Roger," she said in her best Lauren Bacall voice, "You want me to mousse you?" Her Lauren Bacall had a slight east Texas twang.

"Sure, whatever." Roger was thinking about the stock market, which had plummeted 120 points last night. "Just speed it up, I have to get to a meeting."

Undaunted, Nan moussed Roger's hair, her small, Garfield-covered breasts practically in his face.

In the next station, Judy and Lynn, who had been taking in the whole soap opera for the past few minutes, looked on transfixed. It was like driving by an accident – you didn't want to look, but you couldn't help yourself. So far, however, no one needed the Jaws of Life. But it could happen any minute.

Finally, Judy tore herself away from Nan and Roger and studied Lynn's hair. It was all one length, slightly below the tops of her shoulders, thick and wavy – the kind of hair women who have silky straight hair long for. Of course, most women with thick and wavy hair longed for just the opposite.

Lynn sat in Judy's chair with a copy of *In Style* magazine on her lap, opened to a picture of a woman with short, but stylish hair. This was the transformation Judy was going to try to achieve. Deep under the sarcasm and pessimism, Lynn was a trusting soul after all.

Judy picked up a pair of scissors and snipped the perm-scented air for effect. "You're really going to do it? I won't get halfway done and you'll change your mind and run screaming out of here?" she asked.

Lynn gazed at herself in the mirror, a big smile on her face. "Annie likes short hair."

"Annie, Annie, Annie. I've been telling you for years you'd look good with it short, but all Annie has to do is mention it and Voila! you're all for the idea. Now I know where I stand!"

"Cut it out. You know you're my best friend." Lynn looked up to see if Judy was really jealous or just joking. She was betting on just joking.

"Sure, I am. But since I won't have sex with you, it'll be whatever Annie wants from here on. I can see it now." Judy stuck out her lower lip in a mock pout. "Next thing you know, Annie will have you getting a tattoo and learning to ride a Harley."

"Me, a tattoo? You know how I feel about needles," Lynn shivered slightly.

"Love does strange things," Judy said, remembering Mark handcuffed to her bedposts. She couldn't help but smile.

"Who says I'm in love?" Lynn said defensively.

"That stupid grin on your face says it," Judy responded just before she made a four-inch snip on the right side of Lynn's hair. "Well, you can't back out now. From the haircut, I mean. The tattoo, you'll have to go down the street to Think Ink."

As the lock of hair fell to the floor, Lynn inhaled deeply. She wasn't fond of change, even the kind that seemed good on the surface. Even now she had a deep sinking feeling that she'd get her hair cut and Annie would take one look at her, change her mind, and dump her.

"Well, at least with a new do maybe nobody will recognize me and I'll actually get a job. Annie says there's plenty of work in Colorado."

"Oh no you don't!" Judy stopped cutting and stared at her friend. "Besides, what about that job Gail mentioned? Doing PR stuff?"

"I interviewed for it. But I'm not sure I want it even if they do offer it to me. Which I'm sure they won't. After all, I kind of have a reputation to live down. They probably think I'm a wild card who can't be trusted," Lynn said, looking serious and certain, but there was something in her voice that actually sounded hopeful. "Besides, a fresh start in Colorado might be just what I need." She said it like she meant it, but deep in her heart, Lynn knew she was too much of a coward to move now. Before things were really certain between her and Annie.

"You are not leaving me here all alone. I won't let you." Judy harrumphed.

"You're not alone. You've got your boys. And Mark. And Gail. And Donna, April… not to mention your legion of fans."

Judy nodded. "Still. I saw you first. Tell Annie to move here. You said she hardly spends any time at home anyway. So why can't she call Austin home base?"

Lynn had been thinking the same thing. But, as she had reminded herself, she couldn't ask Annie to uproot her whole life and move in with her yet either. It wouldn't be fair to either of them. Plus, she hadn't cleared anything with Coco.

Judy snipped another section of Lynn's hair.

"I only have five weeks of unemployment checks left." Lynn smiled. "Any chance you'll be done with the cut by then?"

Judy poked her with the comb. The phone rang.

"Lone Star Cuts, home of Texas' biggest hair. How can I help you?" Donna answered perkily. Even without asking, Judy knew Donna had had sex last night. A girl can't hide something like that. Judy knew. Donna grinned dreamily at the schedule, "The 17th? Yeah, that works. Uh-huh. I'll tell her."

She hung up and looked over at Judy. "Well, Mrs. Lanier's back. That makes eight stray clients who have returned to the flock."

Judy nodded. Even at her worst moments, she knew in her gut that things would get back to normal if she just waited it out. "They have to come back. I'm the only stylist in town who can mix up Politician's Mistress Blonde #7."

"Yeah, they flock to you like flies to honey. Or maybe that's vinegar, in your case," Lynn grinned.

Judy took another two snips of hair, then glanced up at the clock. 3:45. *Where's Nolan?* she wondered. *He should be here by now.* Her mom alert system had been at Dorothy's Red Shoe Red ever since the day Dick Sloan had practically run over her baby. And the interrogation at school afterward had made her paranoid, although she thought she had a right to be. After all, one of the most powerful men in the state, an asshole who was probably soon going be Governor, had it in for her. Even though he'd won the primary, she wasn't sure he didn't have more tricks up his sleeve to make her life miserable for having mentioned his tiny little problem on stage.

"I wish Nolan would get here," she said out loud.

"You wanna call his school?" Lynn asked.

"Not yet," Judy said as she took another snip. "I don't want to humiliate him like the last three times. He's already threatened to disown me."

Lynn looked down at the hair piling up on the floor below her. "Oh, my God!"

"Just take a deep breath," Judy suggested. "And repeat after me, 'I do not look like a dyke, I do not look...'"

Lynn inhaled. Then exhaled. "I'm okay now." She raised her eyes and watched her reflection in the mirror. "But what if the people at the unemployment office don't recognize me?"

"Maybe you can get one of those big ol' belts with your name engraved in it."

"Yeah, right. That's the style for me. Redneck dyke."

Judy laughed. "Redneck dyke. That's good." The stand-up comic wannabe pushed aside the worried mom. "Let's see… You might be a redneck dyke if you march in the Gay Pride parade and then beat yourself up afterward." She combed hair over Lynn's right ear and trimmed a few stray ends. "Oh, this is good. You might be a redneck dyke if your pick-up truck has a bumper sticker that says '*They'll get my vibrator when they pry it out of my cold, dead hands.*'"

Lynn and Donna both laughed. Judy was unstoppable. "You might be a redneck dyke if you take a break during oral sex to chew tobacco."

"Yuck, Judy. That's disgusting!" Lynn said, laughing anyway.

Suddenly, all hell broke loose at Nan's station. While brushing the hair off Roger's neck and preparing to remove his smock, Nan leaned in and gave him a hickey. It was already turning purple. Shocked, Roger jumped out the chair with Nan practically still attached to his neck, and ran out of the salon, still wearing the smock, a large glob of mousse stuck to his ear. Nan ran out after him, chasing him into the parking lot, past Bernadette's pick-up truck, around the corner past Bobcat Smith's Pet Emporium, where three Labrador puppies watched as if it were a television show produced just for their entertainment, a canine version of *COPS*.

"Roger, come back," Nan yelled. The puppies just stared, their tails smacking against the glass in a rhythmic beat.

Back inside Lone Star Cuts, Donna said, only half-joking, "I don't think she drew blood this time."

"Why doesn't April just fire her?" Lynn asked. "One of these days, she's going to cause real trouble and the police or the Centers for Disease Control are gonna to shut this place down."

Judy smiled. "Hey, you knew Lone Star Cuts was the home for dysfunctional hairdressers when you came in. Remember that waiver you signed?"

With a few more snips, Judy finished Lynn's hair. It had been less than twenty minutes. A new record. "Okay, are you ready?" she asked.

"You're done?" Lynn asked incredulously.

"Hey, give me a break! Occasionally, I get clients out of here before the dust bunnies start gathering." Judy handed Lynn a mirror. "Now it's still wet. It'll shorten up when it's dry."

"I know, I know," Lynn said impatiently, motioning for Judy to swing the chair around, which she did. Lynn stared at the back of her neck. Then she motioned for Judy to let her see the sides. Finally, the front. It was actually very nice. She smiled. "I like it. I really do."

"Yeah, Judy," Donna added, "It's beautiful!"

"You had doubts?" Judy asked in mock anger.

"We're familiar with your work, you know," Donna said.

Lynn was still staring at herself smiling. "I love it. Who knew?"

"I knew, you idiot! I've been telling you to do this for years." Judy picked up the hairdryer and finger-combed Lynn's cropped curls. It took only a minute. "There. Now the girls will be beating down your door. That Annie's gonna have to watch herself."

Suddenly, Nan re-entered through the front door, dragging the smock and Roger's shirt behind her. The crazed look was gone now. She walked to the front desk and glanced around the shop. "So, who's next?"

Donna shook her head. "You've got Vicky Johnson. But not for half an hour." Nan walked away towards the restroom. Donna looked at Judy and Lynn. "We'll have to fit Ms. Johnson with a crucifix first. Maybe give her a couple cloves of garlic." Judy laughed.

Judy removed Lynn's smock and Lynn brushed a few stray hairs from her mint green jacket. "You know Nan needs serious professional help," Lynn offered.

"Don't we all?" Judy said.

Lynn thought about it for a second and nodded. Judy was right. Who could say what normal was, anyway? Hell, just look at the two of them.

Judy dropped her comb into the jar of blue antiseptic and shook out the smock. Lynn got up from the chair and struck a model-like pose in front of the mirror.

"It's beautiful, Judy." Donna nodded enthusiastically.

"I would tip you if I had any money," Lynn lamented.

"That would be a change," Judy said with a shrug.

"I could give you some food stamps," Lynn suggested in jest as she walked towards Donna while hunting for her wallet, which, of course, had made its way into the black hole that all women have at the bottom of their purses.

"I'll add it to your tab. You can pay me when you're a big time reporter for *Texas Monthly*. Or working in the state capitol." Judy added, "The Texas State Capitol, I mean. None of this Colorado bullshit."

"Lynn shook her head. "I'd settle for the *Nickel Ads* right about now. But Dick Sloan's probably got me blacklisted there too. I wouldn't put it past him. Especially now that he's in the race for real."

"He's an asshole. He'll get his," Donna said, handing Lynn her receipt. "Just ask Ima. Karma catches up to us all."

"We can only hope," Lynn said.

Judy walked over and hugged Lynn. Lynn felt bony. Bonier than usual. Maybe stress was taking its toll. Or maybe Judy was hugging with one less layer of fat than the last time they'd embraced. "You'll get a job," she said consolingly. "You have to. How else are you gonna investigate his ass, find out he's wanted in seven states and bring him to his knees?"

Judy noticed Nolan parking his bike on the sidewalk out front. Relief washed over her. Her baby was safe.

Fear is Like Foreplay

Judy sat in her car in the parking lot of the Austin Fitness Center with the air conditioner running full blast, trying to muster up the nerve to go inside. She fingered the free membership card she'd kept on the refrigerator at home, held up with a Pizza Hut magnet, for the past five months. If she didn't use it in the next three weeks – the membership card, not the Pizza Hut number – she'd miss her chance. And with Lynn claiming she had cramps, Judy was on her own exercise-wise.

So she sat, watching the rhythmic motion of people on treadmills and elliptical machines on the other side of the window and trying to decide if she should get out of the car or drive to the Starbucks she'd passed two blocks down. Dan Fogelberg was on the oldies station and she had to at least hear the end of the song before she went anywhere.

Mid-note, there was a knock on her window. Judy looked out, startled. A woman, mid-fifties, more overweight than Judy had been at her heaviest, but with a big smile. Judy rolled down her window.

"Are you okay?" the stranger asked.

"Sure. Just trying to decide if the world is ready for the sight of me in a leotard." She had one on. Under her sweat pants and t-shirt., she but she was pretty sure she wouldn't get the courage to strip down to it.

"Hell, honey, they've seen me in one." She looked at Judy's face. "You're just a skinny little thing next to me." The woman paused and took a deep breath. "I'm Pat."

Judy grinned and turned the radio down. The song was over anyway. "Come on in. It's not that bad," Pat enticed in a melodic voice.

Judy nodded okay, then rolled the window back up, opened the door and got out of the car. "I'm Judy. Judy Parker."

****

Judy had been on the treadmill for sixteen minutes, but it seemed like much longer, more like seventeen. At ten minutes, she had gotten so warm, she removed her t-shirt. At fourteen minutes, the pants came off. Standing in a gym in a purple leotard and black bicycle shorts was the scariest thing she could remember doing. Scarier than childbirth, than stand-up comedy, than telling Juan to get out, than calling his brother to tell him about the accident. Here she was in a leotard in public! And the Earth seemed to still be spinning on its axis.

Pat, sitting on a recumbent bike only a few feet away, was pedaling hard and reading *The Wall Street Journal*. For an overweight older woman, she had great stamina. Her t-shirt, which boldly announced "If You Can't Stand The Heat, Get Out Of My Hotflash," was drenched in sweat. Beneath her shirt, she wore only a pair of red and white striped tights. Horizontally-striped. The woman had stamina and balls. Judy already wanted to be her when she grew up.

"Yessir!" Pat yelled loudly, as she raised her right fist into the sweat-filled air. The other nearby exercisers, including Judy, were jolted from their own thoughts.

"Good news?" Judy asked, panting.

"You betcha! This li'l company I bought last year, they make biodiesel semi trucks, has quadrupled its earnings this quarter."

Judy shook her head. "That's great! Although I have no idea what you're talking about. "

Pat dismounted, folded the paper in half and walked over to show Judy the stock quotes. She tapped her short stubby nail on a company abbreviated ArmBio. "That's it right here. Armadillo BioDiesel. I just made a killing."

Judy stopped walking. Eighteen minutes was close enough to twenty-five in her book. "I always thought that section of the paper was some form of alien communications. 'Turn right at the Milky Way...'"

"It's pretty straightforward once you know what to look for."

"Well, maybe if they had two-for-one coupons," Judy laughed.

Pat smiled and rolled the paper into a tight cylinder. "I didn't know nothin' about this stuff a few years back. But when my Garret died, I decided what the hell?" She smacked the rolled-up newspaper against her thigh. Had her thigh been a snow-covered mountain, avalanches would have been rolling down her leg, covering the villagers who lived on her ankle.

"That's great!" Judy replied as she looked up from Pat's legs to her face. "I mean, the money you're making. I'm sorry about your husband. Were you married long?"

"Thirty-one years." A wistful smile crossed Pat's face.

"That's forever," Judy said, finally catching her breath.

Pat sat down on the bike sidesaddle. Judy dried herself on the dishtowel she'd brought from home. Every other towel in the house was dirty and she didn't trust public towels.

"He had a heart attack one day at the office. His secretary called to give me the news." From the look on her face, it was if Pat was reliving the experience again. Suddenly she slapped her thigh with the newspaper again and came back to the present. "After thirty-one years of being a housewife, I looked around and thought, 'Now what?' Kids were grown, all I had was this huge empty house and spare time. And the business. Garret couldn't get Dan or Tom to take it over, no matter how hard he'd tried, so he left it to me. I guess he figured I'd talk the boys into taking it over once he was gone. But Dan's a teacher and Tom makes furniture, and I just didn't have the heart to guilt them into doing something that wouldn't make them happy. So I figured what the hell, I'll do it! I took some business classes at the community college, and now there's no stopping me. I always thought I wasn't smart enough for business. Turns out I'm smarter than most of 'em. So what if I dropped out of college?" Pat paused. It occurred to her she was being rude by monopolizing the conversation. The thought had not occurred to Judy. "So, what do you do?"

"I'm a hairdresser." Judy sat down on the floor cross-legged. "I always wanted to become a stand-up comic, but..." she trailed off.

"What's stoppin' ya?" Pat asked emphatically.

Judy had just been asking herself that same question last night before she went to sleep. "Oh, the usual. Fear, anxiety, men in trench coats."

Pat laughed. "Well, as I always say, 'Fear is like foreplay.' It's hard to climax without it!"

Judy giggled. "Do you have that on a t-shirt? I'd buy it!"

Both women laughed together. "Foreplay, huh?" Judy repeated. "Maybe you're right. After all, what could happen that'd be worse than spending the whole rest of my life bitching about how I missed my big chance?"

Pat smiled. "That's the spirit. I don't know much about comedy, but I do know that you gotta do what you gotta do. And don't let anyone stop you."

“Maybe I will,” Judy replied, half to herself. It was just seven weeks ‘til the final comedy competition in Houston. And she hadn’t called yet to take herself off the list. Funny how often procrastination pays off.

Just then Pat glanced at her watch, a silver and turquoise thing with huge numbers painted on the dial, she wore because she didn’t have to use her reading glasses to see it. “Whoops, I gotta go. Gotta see some men about a merger.” Pat bounded to her feet with amazing grace for a woman of her heft, and headed toward the door to the ladies’ locker room. “I hope I see you here again! Good luck with your comedy.” With that, she disappeared.

“Thanks,” Judy said to the empty space Pat left behind. “For everything.”

A Man Scorned

There were fourteen unanswered calls from Sloan Campaign Headquarters on Sarah Sloan's answering machine. Twelve on her cell phone. And at last count, nineteen e-mails. Most were from Jimmy Joe. Two were from the head of the Republican Party of Texas. One was from Dick's sister in Chicago.

They were all begging/pleading calls. "It was a long time ago and those things happened back in those days" calls. "You can't do this now, it'll sink the campaign and The Damnocrats will win" calls.

Sarah wasn't planning on answering any of them, but she enjoyed letting them stack up. It showed her just how desperate her soon-to-be ex-husband was getting. Pretty soon he might even be desperate enough to call her himself. That was a call she'd take. There were a few things she wanted to say to Dick, but her lawyer had advised her against calling him and speaking her mind. But if he called her, well, she couldn't be held accountable for what came out of her mouth.

Sarah wasn't the type to sit around and mourn a relationship gone sour, especially one that reeked more than four-month old milk. Besides, she'd been busy. First, she'd called Gail and apologized to her. Dick may have denied Gail's accusations over and over in the media, but Sarah knew they were true. She had been there. And she felt ashamed and disgusted by the whole thing. Sure kids did stupid things in high school, but this was the kind of stupid thing that affected a person's life. She offered Gail money, but Gail turned it down. Dick's winning the primary had disappointed and depressed her, but Gail said getting the secret off her chest and being able to live her life without shame was all she really needed. Besides, it wasn't Sarah who should be doing the apologizing, Gail had said. Sarah told her it'd probably be a cold day in hell before Dick apologized to anyone.

That piece of business taken care of, Sarah had moved on to other projects. Yanking her money out of Dick's slimy hands was a high priority, but that hadn't taken long. She made sure she put it somewhere where he couldn't get to it – nowhere near the joint checking or savings account she had managed since they were newlyweds.

There was one last thing that Sarah wanted to get done. It'd be tough, but somehow she had to find a way to make sure Dick lost the election. She wasn't sure how, but she knew if she set her mind – and her money – to it, she'd get it done.

A Career to Fall Back On

Lynn was having whine with lunch. Again. "The HR guy said he didn't care that I could type 90 words a minute. When he found out I was 'That Lynn Larson,' he just up and left. His own office! He never came back. Had his secretary come back in and tell me the interview was over." Lynn gulped her grape soda. A leather portfolio full of resumes lay on the table between the two friends. "That's how my day went. Thanks so much for asking."

"You can type 90 words a minute?" Judy mumbled between bites of taco salad, from which she carefully removed most of the ground beef and placed it on her napkin.

"Yeah, you know. Career to fall back on and all," Lynn retorted facetiously. "But, of course, that doesn't work if you're persona non grata in this town." She stared past Judy, through the restaurant window at nothing in particular, as if searching for something, but not quite sure what it was. "I guess I'll just have to start looking for a job at a bank or a law firm."

"What about the thing Gail set up?" Judy asked, still chewing.

"Actually, I've got a second interview." Something resembling a smile almost passed across her lips before being banished as premature and presumptuous. "I don't know. Politics is dirty. I mean look at the trouble it got us both in just by telling the truth about one little thing."

Judy laughed. “Now, we’re not sure it’s little. We just know it doesn’t work.”

Lynn shook her head. “That’s not what I meant. Anyway, you know me, I like the truth. I’m not a spin kind of a girl. If they want me to doctor the truth to make it more appealing, I’ll be fired before you can say ‘Jody Foster is still hot.’”

“Give it a chance. Maybe it’s not that bad.”

Lynn shook her head. “No, I’ll leave the lying and exaggerating to the politicians. And the stand-up comics.”

Judy smiled. “I’ll try to hold up my end! Maybe you should go to work at a law firm. If I’m going to do stand-up again, I’m gonna need all the free legal assistance I can get.”

Lynn gazed off into space. Judy checked the space out to make sure there wasn’t something interesting she was missing. There wasn’t. Lynn was just tuning her out.

“I said, I may need free legal assistance.” Still nothing. “You need a man,” Judy offered up as a test balloon.

“Very funny. I heard you the first time.” Lynn looked up. “I was just thinking what a waste of four years of college to end up a secretary somewhere.”

Judy kicked her friend not-too-gently under the table. “Would you just quit that? First of all, this is just temporary, it’ll all blow over, especially once the election’s over.”

“Ya think?” Lynn asked. Then she frowned again. “That’s five and a half months away! I’ll be eating cat food by then!” She paused, thinking about her life and how it had gotten so tangled up in Texas politics. “I never thought I’d be so invested in an election. You know if the asshole gets elected, I’ll have to go to Colorado.” Lynn licked the last few drops of soda off her straw.

"Oh no, you don't!" Judy exclaimed, dropping her fork into her salad. She noticed that tiny beads of sweat had broken out on Lynn's upper lip. Even though it was warm in the restaurant, even with the air conditioner blowing, Judy couldn't remember a time when she'd seen Lynn sweat. "You're not going anywhere. Not unless you take me. And the boys. And Mark. Gail and Glen. Their dogs. Donna. Nan. Leroy. Bernadette. I guess we'll find out how mobile my home really is."

Lynn frowned. "Well, okay, I'll stay for now. But you'll have to find me a job. You know people who know people. Get me in with one of the rich women. Someone who wants to start her own magazine." Lynn stabbed the last bit of fettuccine Alfredo with her fork. Instead of unraveling and falling off as fettuccine does with most people, it clung tightly to the fork until it reached her mouth. Watching Lynn eat was like watching Georgia O'Keefe paint. An artist at work.

"Okay, I'll put 'Find Lynn a job' on my To-Do List, right after 'Win comedy competition,' 'Raise children', and 'Get out of poverty.'" Judy laughed at herself. Lynn was still staring vacantly, only now her eyes looked out across the restaurant toward the restrooms. "Speaking of the competition, are you coming to open mike night Wednesday?" Judy asked, trying to regain her friend's attention. "You know I'm trying out my new stuff."

"Yes, I'll be there" Lynn said, rolling her eyes, "What else have I got to do?" Then, noticing from Judy's face that she may have inadvertently hurt her feelings, she added, "Of course, I'll be there. I'll laugh, I'll applaud, and I'll give you a standing ovation. That way when you're rich and famous, you can hire me to do your PR!"

"I dunno. You'll have to pass the interview first," Judy joked. "Can you tell me why you left your last job?"

Spies R Us

Judy stood in the bright spotlight of Jest Fer Laffs, fingering the microphone cord. "…So, summer's finally here and I was thinking since I've lost so much weight I would splurge and buy myself a new swimsuit. I went to Foley's and this 15-year old perky blonde salesgirl says, 'We have, like, suits, that can, like, hide any flaw.'" Judy did her best Valley Girl with a Texas accent impression. "So I asked her where were the suits that could hide the fact that she's an idiot."

The crowd laughed. The room was three-quarters full. Not bad for open mike night. The manager had already made a note of how many people had come to see Judy. The woman could sure bring in a crowd.

Judy continued. "Not only do we women have to find a suit that hides our flaws, we have to find one that accentuates our best physical feature. So far, I have been unable to find one that shows off my gums to their best advantage."

"Women's bathing suits have to support our breasts, whittle our waists and minimize our hips. The ideal man's bathing suit is one that has a little pocket in front to hold his beer money."

"Speaking of the other sex, what is this fascination Texas men have with football? It's just a game, guys. I swear, if men had periods, tampons would have NFL logos on them. 'Honey, while you're at the store, get me a six-pack of those Kotex Cowboys, super absorbent!'"

Judy paused and took a sip of water. The audience was laughing. She felt good. Gail was right. It was smart to try new material out here before going to Houston.

"I read in a magazine that they've invented a male birth control shot. This makes a lot of sense. Most guys I know won't wear a condom because it's uncomfortable. So, they're going to line up for something that actually hurts? I don't think so. This will probably be as popular as that new microwave treatment for prostate enlargement I read about in the paper. They're still waiting for a man to come plop his testicles in the oven." She paused for effect. "And what setting do you use? Baked potato?"

Judy searched the crowd for Mark's face. He was laughing.

"I say you need to make birth control convenient for men. Mix in it beer. In fact, why not a whole line of spermicidal beer? We could call it Spermweiser. And for those guys whose virility isn't what it used to be, Sperm Lite."

Judy walked over to stage right and grabbed the microphone stand. She dragged it back to center stage and put the microphone in the holder. "Speaking of Sperm Lite, you know the last time I was here I said a few things about Dick Sloan. Well, guess what? The man can't take a joke. I'm thinking if you want to be Governor of Texas, you'd better be able to take a joke."

"You may have heard that Dick's wife left him. Smart lady. Maybe she should run for Governor. She's got all the qualifications: she's smart, she's rich, and she knows how to fake it."

Donna, sitting at the front table on the far left of the stage, shook her head. So much for politically correct. So much for Ima's warnings. Donna glanced quickly around the room. Everyone – Gail, Glen, Lynn, Bernadette, April, Mark and dozens of people whose faces she couldn't make out in the low light – was laughing or smiling. A few were even slapping their thighs. Everyone except for a guy at the very back who looked miserable. *I hope he's not a spy for Dick Sloan*, Donna thought.

He wasn't. He was just one of those guys who enjoy being miserable.

An Offer She Can't Refuse

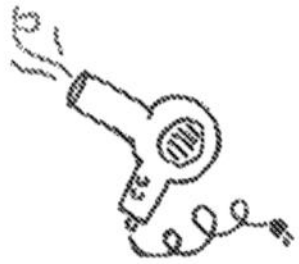

The light bulb over Judy's kitchen table had burnt out two days ago, but she still hadn't gotten around to changing it. To do so meant she would have to drag the step stool out, remove the white globe from the fixture, scowl with disgust at the dust and dead insects that had made their home up there, and then reverse the whole process. And maybe even clean the damned thing off, if she was really ambitious.

So she stood there in the semi-darkness, illuminated only by the light over the sink – a 60 watt bulb trying to do the job of a 100 watt flood light – rinsing dishes.

Despite the domesticity, Judy was actually a million miles away. Or one hundred fifty miles away, in Houston at the comedy competition. She was trying to decide which jokes to keep and which to toss for the big contest. It was only weeks away now and she was getting a little nervous. After all, the prize was $2,000. And, there would be agents in the audience. At least that's what the guy on the phone said when she called to confirm her slot.

As she dipped a plate into the sink, she could hear the TV. She clearly made out the name "Dick Sloan" Almost against her will, she walked into the living room. It was the 6:00 news. Cartoons had long been over. The perky blonde anchor woman with capped teeth and a bright red scarf tucked into her v-neck blue jacket was reading from the monitor. "Today's poll shows Dick Sloan gaining, with incumbent Buddy Lamar's lead narrowing to 11%..."

"Shit!" The dish towel on which Judy had been drying her hands, flew across the room and into the hole in the floor. As if sensing her distress, the phone rang to distract her. On the other end was a voice Judy didn't recognize for a few seconds. It was Sarah Sloan.

"Sarah?" Judy stumbled over her words, "How have you been?"

"Much, much better. Since I left the prick." Sarah sounded a little tipsy. Judy could hear bar noise in the background.

"I'm glad for you," Judy replied.

There was silence. One, two, three seconds passed. Finally, just as Judy was about to interrupt, even though she had no clue what to say next, Sarah spoke up again. "I never told you how funny I thought your jokes about Dick were. You know, the ones they quoted in the paper."

Thanks," was all Judy could come up with. *Yeah, that was witty and memorable.*

"Judy, I know you don't know me very well, but I feel like there's a bond between us…" Sarah's words trailed off. "Well, I should just tell you why I called."

"Only if you really want to. I'm just glad you're not trying to sell me something." Judy finally relaxed. Hell, it was just another Lone Star Cuts client. So what if no one as rich and powerful as Sarah had ever called her at home. So what if Judy had tried to ruin her husband's, make that ex-husband's, career.

"Oh, but I am trying to sell you something." Sarah sounded mysterious and suddenly more sober. "I have this great idea. At least I think it's great. And so do a lot of my friends at the country club. Anyway, I'm calling to invite you lunch. I have a proposition for you."

"A proposition, huh? You're sure you want me and not my friend, Lynn. You know, she's the lesbian."

"Not that kind of a proposition," Sarah laughed.

"Oh, good," Judy said in mock relief. "I haven't given up on men yet." She looked at the clock, reminding herself that Mark would be there soon.

"Me either. I've just given up on assholes. Of both genders." Sarah sounded like she was getting further and further away from the receiver.

"I can hardly hear you," Judy said.

"Sorry!" Sarah's voice got louder. Almost too loud. "What I'd like is for you to come to lunch with me. To talk about my proposition. I think it would be lots of fun." There was another lengthy pause. The best part of my proposition is that we get to nail Dick to the wall!"

"That's all I need to hear," Judy said enthusiastically. "Count me in. I'll bring a hammer! Where? When?"

"Tomorrow. The downtown Hilton. Is 2:00 okay?"

"It's great. Tomorrow's my day off."

"Okay, see you there. And Judy, keep your mind open," Sarah said. Then she hung up.

"I wonder what that's all about." Judy asked no one. But then she heard a key in her lock and smiled. Mark had a key to her place and it felt right. She ran to the door – all seven steps – and greeted him with a hug.

"You'll never believe who just called."

Breadsticks Make Strange Bedfellows

Enriched wheat flour (contains niacin, reduced iron, thiamine mononitrate, riboflavin, folic acid) vegetable shortening (partially hydrogenated soybean oil), salt, baking soda, malted barley flour, calcium carbonate, yeast. 50 calories per serving, total fat 1.5 g.

Judy sat alone at the table, munching on a bread stick, reading the label. Not for individual sale. *I wonder what would happen to me if I tried to sell these on the street.* she thought. *Would the Breadstick police come after me with handcuffs? Maybe I'd have a yeast infection for the rest of my life.* Judy laughed to herself.

She glanced at the clock on the wall above the doorway separating the restaurant from the bar. 2:13. *Maybe I shouldn't have let the waiter seat me. I doubt Sarah will see me way back here*, Judy wondered. An amusing thing to wonder, considering she was wearing a lime green dress that apparently had been knit from some phosphorescent material manufactured from radioactive waste. No one looking could have missed her, even in the dark.

Just then, Judy spotted that unmistakable red hair across the room. She waved to get Sarah's attention.

Sarah walked quickly to Judy's table. Heads turned everywhere along her path. Sarah was stunning in her navy pantsuit and yellow silk blouse. As she pulled her chair out and started to sit, Judy noticed immediately that years had melted from Sarah's face. She also noticed the missing wedding ring had been replaced by a tactful, but large, emerald ring.

"Hi," Judy said nervously.

"Hi," Sarah answered back. "I'm so glad you could come." She picked up a menu. "Have you decided what you want?"

"Yeah. I want the grilled tuna on tomato and I want to know how we're going to nail your ex to the wall," Judy said unable to contain her curiosity any longer. "Oh, and I want world peace and fat-free chocolate."

Sarah smiled. Their waiter came over to the table and asked if they were ready to order. Sarah said yes. "We'll both have the grilled tuna. What do you want to drink, Judy?"

"Dr. Pepper."

"One Dr. Pepper and one bourbon and water." She handed the waiter the menus and he walked off. "Okay, let's talk turkey." Sarah folded her hands together and looked Judy straight in the eye. "Here it is. Now remember, you promised to keep an open mind." She paused. Judy couldn't tell if she was trying to decide how to say what she was going to say or if she was just going for effect. "Judy, I think you should run for Governor."

Judy almost swallowed her tongue with surprise. Then she started laughing so hard the water glasses on the table shook. "Yeah, right. Okay, good joke. Now why am I really here?"

"Really. Let me explain." Judy noticed that suddenly the restaurant seemed totally silent. She could hear herself breathing. She could hear her heart. She could hear her stomach trying to digest the riboflavin and partially hydrogenated soybean oil from the breadsticks. What the hell was riboflavin anyway?

And somewhere in the background she could hear Sarah's voice, far, far away. "Dick is an asshole. I've known this for a long, long time. I tried to ignore it because I was too lazy to leave him. But now I'm facing it. He slept around on me all the time. He date-raped that woman in high school. And there's that woman from the bank he managed who says he sexually harassed her…"

Judy broke in. "Let's not forget his habit of running over small children and then driving off like a bat out of hell."

"Exactly." Sarah slammed her palm on the table. "I'm sorry."

"That wasn't your fault. None of it was your fault." Judy picked up another breadstick. She looked down at the table and noticed that at some point while she had been busy listening to her bodily functions and trying to hear Sarah out, the waiter had brought their drinks to the table. She took a sip. She felt a little better.

"My ex is not the kind of man who should be in charge of this state. But it looks like that might happen unless we do something." Sarah sipped her drink through the tiny straw. The ice cubes clinked in her glass. *Maybe she's still drunk from last night*, Judy thought. "I've thought about it a lot," Sarah continued. "And I think you're the answer."

"Me? I think maybe you've had one dye job too many. Those chemicals have affected your brain," Judy laughed.

Sarah ran her finger across the top of her glass. "For weeks now, I've been trying to think of a way to make sure that Dick gets what he deserves. I thought about putting arsenic in his barbecue, sending naked pictures of him into *Livestock Illustrated...*"

"Oooh. That wouldn't be fair to the livestock," Judy interrupted, giggling.

"So I started talking to people. Well, women mostly. Rich, influential women." She paused. "And we all agree that if we could find someone to run as an independent candidate for Governor, someone who doesn't stand a chance to actually win, but who would have the guts to tell the truth about my ex and who could do it in a way that would make other women stand up and pay attention, there's a chance we could swing enough votes away from Dick so that he'd lose."

For once in her life, Judy was speechless. As she sat there dumbfounded, staring blankly at Sarah, who was now just a fuzzy blue and yellow image with an orange head, the waiter brought their orders. Sarah thanked him and he left without even noticing the catatonic state Judy was now in.

Sarah pressed on. "You're funny."

Judy nodded from her mental fog. *That's true. But what did it have to do with running for Governor?*

"And everyone likes you." *Also true*, Judy acknowledged to herself. *But that didn't mean they'd be foolish enough to vote for her.*

Sarah wasn't finished. "And I know you have balls, you have to have balls to do comedy." Sarah paused. "And, as an added bonus, you're not too fond of Dick, right?"

Judy sat up straight in her chair. She noticed that her right hand was clutching her napkin so hard there was no blood in her fingers. "You're right about that." Judy looked around the restaurant. Things had come back into view. "Not to mention that my best friend is threatening to move away if he gets elected because of what he'll do to gays and lesbians."

"That would be Lynn, right?" Sarah asked as she smoothed her napkin on her lap and then began to cut into her tomato.

"Yes." Judy tapped a breadstick unconsciously against the glass table top. "Okay. I'm funny, well-liked, ballsy, and I hate your husband. I don't think those are the traditional requirements to run for Governor."

"No, they're not," Sarah agreed. "But who says we can't break with tradition?"

Judy started eating. And thinking about it. And thinking she was crazy to be thinking about it. And thinking, what the hell, everyone knew she was crazy anyway.

Sarah pressed on. "Think of it as a huge stand-up comedy show. You'd write lots of jokes about Dick and about the election. And you'd do them on TV." Judy's ears perked up. TV, huh? That could really jump-start her comedy career. "My PR firm would do all the publicity. And my friends and I would pay for the whole campaign."

Boy, that Sarah really knew how to proposition a girl, Judy thought. "You've got this all thought out, huh? But isn't it too late to start running?"

"All we need is 5,000 signatures." Sarah finished up her drink and motioned to the waiter for another. "With your friends and my friends, we could get 5,000 signatures in our sleep."

"Yeah. I know a lot of people. Hell, Nan alone is seventeen, eighteen…"

They both laughed. "Wait a minute. There's a problem," Judy said.

"What?"

"I've got this comedy competition in four weeks and six days. In Houston."

"That's not a problem. It's an opportunity. We'll work with it."

"Okay." Judy sat quietly for a few seconds. Then an idea popped into her head. "You know, Lynn has been looking for work ever since Dick got her fired. She's a journalist. We'd have to make her part of the campaign."

"Sure, she could be your campaign manager if you want." Sarah folded her napkin. Her lunch was only half-eaten. "Actually, that's a great idea because she really knows you better than I do. Not to mention that it would really cook Dick's goat if he went down in flames because of someone whose campaign was managed by a lesbian. He hates lesbians. Finds them a personal affront to his masculinity."

For the first time since Sarah sat down, Judy looked her straight in the eye. She was no longer feeling like a second class citizen, a high school drop-out, or a woman on the verge of bankruptcy every time the rent was due. "I have another idea. If we're trying to get women to vote, I think a great way to get to them is through beauty shops. We could run the campaign from Lone Star Cuts and do promotions in beauty shops throughout the state."

Sarah looked genuinely interested in the idea. "See, and you thought you couldn't run for Governor! Already you've come up with two ideas that are better than what I had in mind."

Judy took a deep breath. What was she thinking? This wasn't going to happen. "Hey, this is a great fantasy, but it would never work. I'm a hairdresser. Did you know that I didn't even graduate from high school?" She stopped for a moment. Of course Sarah knew. It was in the paper.

"Who cares? Besides, if a high school drop-out hairdressing comedian and a lesbian cause Dick to lose the election, what's left of his boy parts will shrink up completely. It will be so satisfying."

"But what I know about politics would fit..." she said, but was interrupted.

"Hell, I know a thing or two. How do you think Dick got where he is today? You don't know this about him, but he's just as stupid as he is mean. If it weren't for me and J.J., his campaign manager, he'd be shoveling cow patties in Lubbock."

"I'm not very smart either," Judy said quietly.

"What do you mean? From what I can tell you're one of the smartest people I've met. You're a single mother raising two sons on a hairdresser's salary. That takes a hell of a lot more smarts than running for Governor of Texas." Sarah said it so matter-of-factly Judy almost believed her.

Judy laughed. "Fear is like foreplay," she said under her breath.

"What?" Sarah queried.

"Oh, just my new catch phrase. 'Fear is like foreplay.' And this idea scares the hell out of me. So maybe it'll be a hell of an orgasm at the end."

Sarah laughed. "Between us we've got the money, the friends and the balls to do whatever we want. Besides, I think this will be more fun than using Nads to remove Dick's pubic hairs."

Judy finished the thought, "Maybe we can do that after we trounce him at the polls."

The two women stood up and hugged. Who says all the big deals are made in the men's room?

Sarah was just about the leave the table, when Judy had another thought. "I don't know how you feel about having a man in the campaign, but my boyfriend Mark works for Richards & Romano Polling. I'm sure there's something he could…"

Sarah turned around. "Judy, that's just perfect. This may just work!"

Hair Today, Governor Tomorrow?

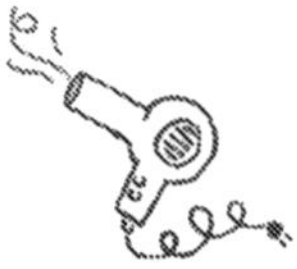

This is crazy! Judy thought for at least the four thousand three hundred and nineteenth time today as she shook yet another customer's hand, trying not to trip over the rented folding chairs that were crammed in the place. *I don't know who's more insane, me or Sarah.*

Judy took a second to catch her breath and looked around. The salon was already filled with regulars and a few people she didn't know. Most were women, except for Leon and Glen and Buster and two guys she didn't recognize. Sarah and four of her rich friends were standing near Donna's desk drinking coffee and looking at the giant hairspray bottles.

Deciding it was time to bring the meeting to order, Lynn stood on a chair and whistled through her fingers. It was a trick she learned from her brother when she was in the third grade. Who knew what life skills would come in handy down the road? Slowly, the crowd moseyed towards the chairs. Half of them ended up standing against the walls and leaning on the counters.

"Welcome to the preliminary organizational meeting of the Judy Parker for Governor campaign," Lynn said with enthusiasm. Judy smiled at her. At first Lynn had thought this idea was as kooky as Judy had, but being campaign manager for Judy's "Comedy Campaign," as Lynn referred to it, had given her purpose. She hadn't scowled in over a day. The audience broke into applause.

"We're glad you could all be here. Since we don't have much time to get this campaign up and running, I'm hoping we can get a lot done tonight." There were nods and murmurs in the crowd. "Okay, our first order of business is coming up with a theme for Judy's campaign. Any suggestions?"

Just as three hands shot up, Mark came in the door carrying a thick file. Most of the women turned around at the sound of the chimes to see who else was joining them in their subversive little scheme to take over Texas politics.

Judy smiled at Mark. *Wow, she thought. He's really going to do this with me.* She already knew she loved him, but here was a man who was not only willing to put up with her crazy ideas, he was there to help steer the boat on the trip to Looneyville. She never expected she'd find anyone like him.

Bernadette, whose hand had gone up first, spoke up. "How about 'What do women want? Well, if you'd just shut up and listen for once, we'd tell you.'"

Most of the women laughed. But Lynn shook her head. "It's a little long. Besides we don't want to alienate the male vote. A few of them might even vote for Judy."

Mark yelled from the back of the room, "You got that right!"

Sarah nodded her head. "Yes, let's not assume only women will vote for Judy." Again, the group snickered. "I think the slogan should play to Judy's strength, which is her biting humor," Sarah added.

One of Judy's regular clients added, "Yeah, we all know it's not her styling talent."

Judy folded her arms over her chest and said, "Thanks a lot! We'll just see what color your hair turns out next time." Suddenly an inspiration hit. "How about 'At Least She's Only Irrational Once a Month?'"

"I like it," Gail said laughing.

"I've got another one," Judy added. "She's Less Annoying Than a Yeast Infection."

Everyone laughed. "Funny, but it won't play in Denton," one of Sarah's rich friends said in a husky voice.

"Yeah, I forgot women in Denton don't get yeast infections," Judy joked. "They pay someone else to get them for them!"

Lynn scribbled something down on her pad. "How about 'There's a Big Heart Under That Big Hair'"?

Someone else spoke out. "I like Judy's first one best."

Judy, though, thought she could outdo herself. "What about 'Let's Make a Permanent Wave?'"

Everyone got that look, the one that says, *Hey, that's genius*, and smiled and nodded. Lynn quickly took a vote and the campaign had its slogan.

"Okay, we've got our slogan." Lynn pressed on. "Now, how's the signature-gathering going?"

From somewhere in the middle of the crowd, Maddy Lou stood up, looking exceptionally tan and wealthy. Through half-glasses perched on her nose, she looked down at her notes. "We're up to about 3,783 signatures. Judy, your idea of collecting signatures in beauty salons was a stroke of genius! We just tell 'em a beautician's running for Governor and they can't wait to sign up."

"Maybe they can't hear you over the hair dryer and they think they're signing up for a date with Ashton Kutcher," Judy suggested.

"No, Judy," Maddy Lou added. "This was really a great idea. I think women are really fed up with the yahoos who've been running the state for the past thirty years, except of course for my personal friend Ann Richards, God rest her soul and her big hair too."

"Amen," someone said.

"I did her hair once," Judy exclaimed.

"We know," Donna said from her seat in the front row. "I personally have heard that story eighty or ninety times."

"Do you need any more help gathering signatures?" Lynn asked. Maddy Lou shook her head. "I'd say we'll have it done in three days." She sat down.

"Okay, great!" Lynn said. She placed a check next to "Signatures" on her agenda. "Great job, Maddy Lou. You too, Ima, Bernadette, Leon, Sally, April..." Lynn paused.

"April," Judy spoke up, "Thanks for making those phone calls to salon owners. That really helped a lot."

"Glad I could help," April still couldn't get over the fact that Judy was actually running for Governor of the state. That just blew her away. And it reinforced her belief that even though she had a tendency to choose dysfunctional hairdressers, they were mostly good people.

"Yeah, and thanks for letting us use Lone Star Cuts as our headquarters too," Lynn added. "I love that sign outside: 'Lone Star Cuts – Home of Texas' Biggest Hair and Our Next Governor.'"

"That was Sarah's doing. I just hung it up."

Sarah smiled from the back. It was great to be working on a campaign where people actually gave each other credit.

Lynn kept going. "Okay, next. I've contacted the few friends I have left in the media and we should start seeing coverage of Judy's campaign tomorrow. *The Austin American Statesmen* is running an interview with Judy tomorrow morning. It's in the City-Region section. They'll treat it as a joke, but we all know that already. The idea is to keep the joke going long enough that it makes people think before they vote." She paused and looked around. It felt good to be doing something important again. Especially now that she was earning a paycheck again. Sarah had given her a nice check for signing on as Judy's campaign manager.

"And Wednesday, Channel 3 will send a reporter to interview Judy and, Sarah, I said you'd be there too. I hope that's okay. It sealed the deal."

"No problem, Lynn," Sarah drawled. "Whatever it takes."

"Excellent. If we get enough media attention, they'll have to let Judy in a debate or two. And that's what we're counting on. I know I'd hate to be in a battle of wits with her, so Dick will go down like the Titanic if he has to face off with our girl here."

A cheer went up from the crowd. Judy would have blushed if it was in her, but she wasn't the blushing kind. Besides, she knew for a fact, that even if she didn't know as much about the issues as the "real" candidates, she was definitely more entertaining. And TV was really all about entertaining the audience. Furthermore, Sarah had hired a tutor for her to catch her up on important things like highway construction, prison statistics, and pollution of the aquifer system. Most nights lately, Judy had gone to bed with her head swimming with things she'd never paid attention to in her life. And she had gone to bed with Mark. He hadn't exactly moved in – after all, she did live in a trailer. And Sarah thought that was a good sales point – single mom living in a trailer doing hair for a living. It was the kind of thing a lot of women voters in the state could relate to.

Lynn moved onto the next agenda item. "Next, Mark will share the latest poll numbers with us."

Mark came up to the front of the room to the whistles and applause of the crowd. Judy kissed him playfully.

"Hey, I have to remain unbiased. So no kisses from the candidate. At least in public." Mark said with a smile. Judy smiled back. What guy in his right mind would help his girlfriend run a mock gubernatorial campaign? He must be as nuts as she was.

"The good news is," Mark began, "Dick Sloan has already slipped a little in the polls in the past week."

A roar went up from the crowd.

"So we need to keep the ball rolling forward. Once Judy's campaign starts being taken seriously, or comedically, our goal will be to pull 4-5% of the voters from the Sloan campaign, without pulling any from the Lamar campaign." Mark tapped his pen, like he was all business.

"Thanks Mark," Lynn said.

"Thanks honey," Judy whispered in his ear.

Mark smiled. "I gotta get back to the office. I'm here on my break." He made his way through the crowded rows of chairs and headed toward the door. "Good luck!" he yelled over his shoulder to the whole room.

Lynn plunged ahead, trying to make sure business got done. "Okay, Judy, you need to make sure when you're talking to the media that you focus your wicked sense of humor on bashing Dick, not on the entire electoral process. Got that?"

"Oh you bet!"

"Good," Lynn said.

Sarah waved her hand from the back of the room. "I think we need a list of things Judy needs to NOT talk about during the campaign. You know for those women in Denton. Like yeast infections and tampons."

"Vibrators," someone piped up.

"Ménage au trois," another voice countered.

"Oh, come on now, you're taking all the fun out this." It was Judy this time.

"I'll make a list," Sarah concluded.

Lynn nodded in agreement. "Wouldn't want to unleash Judy on the world of Texas politics without some kind of guidelines, would we?" She paused for a moment. "Okay, let's work on our platform."

"Platform number one, no platform shoes," Judy yelled out. "Been there, done that."

Thelma and Louise and Louise

They drove with the top down, even though Judy warned them of impending hair doom. Convertible hair was worse than motorcycle helmet hair, she'd said. Judy wore a scarf to protect hers. After all, she had to look good now that she was in the public eye. Constantly, it seemed.

Gail drove 78 m.p.h. in a 65 zone, but no one was complaining. That's why she was driving. She was fast but safe, and for some reason always seemed to know where the cops were just in time to slow down. Judy said she must have ESTP, extrasensory speed trap perception.

It was Gail's idea to turn off the A/C and put the top down. They had rented a convertible Thunderbird just for this trip, something Gail had always wanted to do, but had never had the chance to. With the dogs in the back seat, she was afraid of permanent hearing damage or bugs up their snouts.

Lynn had agreed, only because she was in such a good mood lately, she was saying yes to all kinds of things. Being in love and feeling needed snapped her out of her depression faster than a gallon of ice cream. Besides, now with her short do, Lynn didn't have to worry about what the wind was doing to her hair.

Judy sat in the middle between Gail and Lynn. Even with the wind rushing past and having to scream to be heard, for the first time in her life, Judy truly felt like part of a family. These women really adored her and would do anything for her. Sure, she had her boys and loved them to death, but if it came down to choosing between saving her life and saving their PlayStation or their rats, Judy was pretty sure which choice they would make. Judy smiled with her mouth open, even though she chanced swallowing a Texas-sized mosquito in the process.

They were on their way to the Houston Comedy Competition. Despite the fact that Judy had faced interviewers, TV cameras, and political underlings from Dick Sloan's campaign and had stared them all down fearlessly, she was nervous. She could feel a trickle of sweat running down the back of her neck and settling in the elastic of her new panties. When Sarah had taken her shopping for appropriate campaign clothing, she'd bought all new undies too. At Victoria's Secret no less. Judy's old undies had come from K-Mart.

Lynn and Gail had tried to convince Judy she'd do just fine, that she was the funniest person they'd ever met. But this was Houston. There were like three million people in Houston. *A lot of them must be funny too,* Judy thought, biting the inside of her lower lip. Not to mention the other two comics from Austin. And, there'd be people there from other big cities like San Antonio and Fort Worth and Dallas. They had to be funny too. *Well, not Dallas*, she thought. Judy couldn't imagine anyone from Dallas with a sense of humor.

Judy was also nervous because for the first time since her campaign started, Mark wasn't going to be there to support her. She had started to count on his calming influence and his kisses afterward. But Mark had a job to do, and with the campaigns in high gear, he was putting in overtime at the office. His boss had been making threatening noises about how close he was to her campaign, so Mark and Judy had agreed it was probably best for them to lay low for a while.

It's better that he's not here, she thought. *I don't want to have him see me die on stage.*

To distract her from her worries, Lynn went over a checklist of things yet undone. Lynn didn't miss the irony of trying to take Judy's mind off worrying about doing stand-up comedy by reminding her of her obligations as a gubernatorial candidate. She had to yell each item out, but fortunately had everything printed out in a binder, so if Judy didn't hear, she just pointed with her pen. Lynn jotted down a note reminding herself to call to make sure the meeting with the Texas Women's Medical Association was confirmed.

The traffic got heavier as they reached the outskirts of Houston. Gail noticed a lot more expensive cars now too. Big oil and big bucks. That's what Houston was about. *I could never live here*, Gail thought, as she came almost to a standstill in the late afternoon traffic. *The dogs would hate it.*

"We're almost there. What was the address again?" Gail said in a nearly normal voice now that they were practically at a standstill.

Judy pulled her confirmation out of her purse and read the address out loud. "12255 Richmond. The club is called Giggles."

"I know where it is," Lynn said as she filed away her notebook and ran her fingers through her hair, making it spectacular once again. "We've got three hours 'til you've got to check in. What do you say we go to Galleria?"

Judy nodded enthusiastically. "That'd be great! Shopping would definitely take my mind off the competition."

Lynn shook her head. "That's not what I had in mind. There are two swanky beauty salons there. I think you should go shake a few hands, comb a few heads."

"Of course, I should have known you and Sarah wouldn't give me a second's rest." Judy looked imploringly at Gail. "Help me. I have no control over my life!"

Gail patted her friend's knee. She wasn't using her hands to steer anyway since traffic seemed to be backed up for at least the next two exits. "It'll all be over in November. Besides, you love it and you know it."

Judy sank her head into her hands and shook her head from side to side. "I know. But I'd like to have just one minute to do something not related to running for Governor. Get my nails done. Buy a blouse. Have sex."

Gail laughed. "Well, we can't help you with the latter, although I won't speak for Lynn."

Lynn smiled. "Sorry, I'm spoken for." She took out her Blackberry and reviewed the day's schedule. "O.K., Judy, I'll tell you what. If we can hit both salons in an hour and a half, we can squeeze in some shopping. Deal?"

"Deal!" Judy said, sitting up straight.

Gail found a slight gap in between two cars and pulled over. "That's assuming we ever get off this freeway!"

Warning: This Chapter is Shorter than It Appears

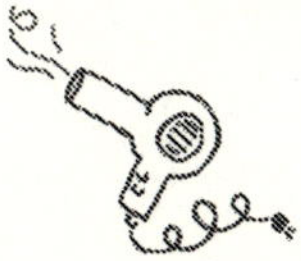

Judy stood on the Houston stage looking disappointed. Third place! She thought she deserved at least second. Gail and Lynn thought she should have won. But, they were slightly biased.

Sarah, who had flown in on her private jet, which would take Judy to El Paso at midnight, was not in the audience. She was, instead, waiting for Judy at the TV station forty-five minutes away for their interview.

Oh, what the hell? Judy thought. *I may not have won, but, I am the only comic here running for Governor. That counts for something. And $800 is $800. When this whole campaign thing is over and I'm back to living in my trailer and doing hair for a living, $800 is going to come in real handy.*

The Problem with Uppity Women and Exes

Jimmy Joe wandered back and forth, hoping that he could sleepwalk through the rest of the campaign and then get on with the rest of his life. Unfortunately, Dick came through the front doors, talking on his cell phone and putting an end to Jimmy Joe's trance. Dick snapped the phone together as hard as he could.

"Damn! That was Ronald. According to the latest poll, we've lost another two percent!" He shook his head in disbelief. "I can't believe this stupid Judy Parker thing may blow us out of the water!" He looked around the room. "Hell, she does hair for a living. And she's a high school drop-out."

Jimmy Joe looked at Dick. He seemed smaller than usual. But no less mean. "Well, she was smart enough to get your wife to underwrite her campaign." He turned his back on Dick and muttered under his breath, "Which makes her smarter than you."

Dick slumped down into an orange plastic chair with cigarette burns in the arms. "We're still gonna win this thing, it's just gonna be harder than we thought. But if Arnold Schwarzenegger can get elected after groping all those Hollywood gals…" He stopped mid-sentence, as if trying to muster up a little of the raw appeal the former Mr. Universe had had when he'd run in California. But Dick was no Arnold. Although some had said Arnold could be quite a dick.

Jimmy Joe couldn't hold his disdain any longer. "You bet your ass it's gonna be hard. The numbers show that a lot of the women who would have voted for you even after the whole date rape story, even after Sarah started telling people about your private life, are thinking about voting for Judy Parker. And after that fiasco at the Texas A&M commencement where you called everyone who hadn't gotten a college degree 'an idiot,' you lost a lot of blue collar votes as well."

Dick stared at him blankly for a few seconds, as if the idea of losing the women's vote and the blue collar vote was supposed to mean something, but somehow it didn't. After all, he still had the country club vote, the oil vote, the lobbyist vote, the hunters' vote. In other words, all the votes that really counted in Dick's book. "Screw the women. Screw the blue collar fuckers."

"You know what, Dick? Screw you!" Jimmy Joe said with as much vitriol as he could muster in his exhausted state. "I am sick and tired of running interference for you. Painting a nice face on a pile of crap. You think this is so easy, try doing it for yourself. I quit!" Jimmy Joe found his car keys in his pants pocket and quickly headed for the door.

As it closed, Dick continued to stare blankly, not realizing that not all the shit had hit the fan yet and now there was no one left to wipe it off.

Only Her Lt. Governor Knows For Sure

"Shhh! It's coming on!" Donna whispered excitedly as she pressed the volume button on the remote. All the stylists, with the exception of Nan, who was MIA again, and Judy who was in San Antonio campaigning, had gathered around the couch in the back room of Lone Star Cuts to watch Judy's newest commercial.

One the screen, little cartoon men in suits started pounding their fists and butting their chests. A giant hairdryer appeared and blew them away. Then Judy's face came on the screen and an announcer said, "This woman thinks she knows how to make things right in Texas government. Does she? Or doesn't she?"

"Hi, my name is Judy Parker, and I'm running for Governor of Texas." Judy looked great. She was wearing a hot pink suit Sarah had picked out for her. And all the running around campaigning had helped her drop another fifteen pounds. The Campaign Diet, that's what Judy called it. Maybe once this whole thing was over, she could think of a way to sell it.

"Now you're probably a little skeptical – I bet you even have a few questions," Judy continued on the commercial. It cut to woman wearing a smock that said Ilene's House of Beauty. "Judy," she said, "do you really think you can win this election?"

Cut back to Judy: "Good question! Actually, I think my chances of winning this election are slimmer than Dick Sloan's chances of being elected president of the National Organization for Women."

Cut to second woman standing on the street next to the state capitol building, sporting really big hair: "So your campaign is just a big joke?"

Cut to Judy: "Well, when you stop to think about it, wasn't the gubernatorial – I hope I pronounced that right, it's a big word for a high school drop-out – race a big joke before I got into it?"

Cut to third woman, sitting in a mammography waiting area. "If you don't think you could win, why are you running?"

Cut to Judy: "I have three goals in my campaign for Governor – well, four if you count getting out of my real job for a few weeks. My first goal is to get more women out to vote. Now, I'll be the first to admit that in the past, I would rather get a bikini wax than pull a lever for most of the yahoos that are on the ballot. But that's the problem. As long as we women don't vote, the yahoos keep coming back."

"I want you to take one look at me and say, 'Look at her! She's a high school drop-out! She cuts hair for a living! But she took the time to run for Governor, so I could at least take the time to cast my vote!'"

Judy's monologue was temporarily disrupted by a spontaneous burst of applause from Donna and Bernadette's "Give 'em hell, Judy!"

"My second goal is to get the yahoos to pay attention to us. When it comes to what women want in this state, most of our politicians have attention deficit disorder. Their minds wander. They start thinking about the last Cowboys game or whether there are any long necks in the fridge."

"Men are always asking "What do women want?" In my campaign, I plan to answer that question. We have some real serious issues that no one seems to want to deal with."

"And my last goal is to make sure that no matter what happens in this election, Dick Sloan doesn't become Governor. We're talking about a man who date-raped a friend of mine in high school. And he's shown no remorse, made no apologies. I'm sick and tired of being screwed by politicians. So let's not elect one with a proven record of doing that over and over."

Cut to graphic: "Big Haired Women Unite: Let's Make a Permanent Wave." Judy's voice over: "Come on girls, let's make a permanent wave."

As the picture of Judy faded out, the TV voice-over announcer said in that ever-so-professional TV voice-over announcer voice: "Sarah Sloan, Treasurer. Yeah, she's voting for Judy too."

The room went wild.

How Can We See Eye to Eye When You're Staring At My Breasts?

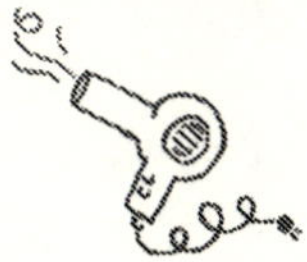

Judy shivered. *Why is it always so cold in television studios?* she wondered, as she rubbed her hands up and down her arms to warm up. Her silk teal pantsuit looked great on TV – she'd worn it in Wichita and Abilene – but it didn't do much to keep her warm.

Judy peeked her head out the green room door and down the hall. She could see the audience filing in. Connie, the League of Women Voters representative who had met her in the parking lot and showed her the studio, was now helping show the audience to their seats. Judy returned to the green room, which was actually blue, sat down on the worn sofa, and glanced over her notes as she sipped her Dr. Pepper.

Quietly, the door opened. So quietly, in fact, that Judy didn't even look up from her notes.

"Are you ready?" Gail asked.

"Where'd you come from?" Judy asked happily as she went over to hug her friend. Judy was careful not to smear her make-up. "Sure, I guess I'm ready. After all, only a couple million people are gonna be watching."

"Just think of it as a rehearsal for your first HBO special." Gail looked as calm as Judy wished she felt.

"Well, let's just hope they have a sense of humor," Judy said with a half-laugh.

"You'll kill," Gail reassured her friend.

"We can only hope. If not, there's always Big Hair and Big Bazooms." Judy walked over and glanced at herself in the mirror.

Gail winked at her. "You look great. You'll do great." She glanced at her watch. "I'd better get out there so I can get a good seat." She squeezed Judy's arm and left as quietly as she came in.

Judy turned away from the mirror to the far wall of the green room. It was adorned with photos of local celebrities who had been guests of various shows at the station. As she leaned in to the check the signature on one, the door opened again.

Dick Sloan walked in, looking down at his note cards. He didn't notice Judy in the corner. She, however, had heard the door this time, turned around, and stood quietly watching him, trying to brace herself for their first face-to-face meeting. Finally, he looked up.

Staring first at her breasts, as Dick always did when confronted with a woman, he fumed, "Shit! What are you doing here?" he growled.

"Looking for a date!" she responded facetiously. "Whaddya think? I'm here for the same reason you are."

He dropped several of his cards on the floor and bent down to pick them up. "Shit. The, Ladies' Voting Society is..."

"I think you mean the League of Women Voters." Judy corrected him. "I've hear they get sore if you get the name wrong. They could all be PMSing at the same time, you know how we women are." She laughed. This was going better than she had hoped.

He glared at her. "They're only supposed to invite legitimate candidates to this debate."

"I'll have you know my parents were married before I was born. Though not by much." Judy would have continued, but Dick had opened the door, stepped out and was yelling down the hall.

"Jimmy Joe! Jimmy Joe, get the hell in here NOW" The hall reverberated with the sounds of silence. Dick returned dejectedly. He refused to believe Jimmy Joe had really left him, even though it had been over a month now. He'd managed to escape his new campaign manager, Roger – a young Republican hotshot the Texas Republican Committee had sent in as a replacement to make sure Dick didn't make any other major screw-ups between now and the election – by faking a phone call and escaping out the back of the restaurant at which they'd been prepping. Dick hated Roger because he refused to take his abuse and had his own ideas about how things were going to be done. And since Roger had the power of money behind him, Dick was forced out of his alpha male position. He didn't like that one bit.

Dick sat back down on the sofa, as far away from Judy as possible.

"Looks like you're on your own," she said. "What's the matter, Dick, you scared of me? I promise, I'm unarmed. Look, no scissors." She was actually feeling a little sorry for him. He was going to have to go on TV with that face. No amount of make-up could make that appealing. Not to mention the hair.

Mostly under his breath, he stammered, "How the hell did you talk 'em into letting you on tonight?" She didn't answer. Then it came to him. "Shit! Sarah. Damn her!"

Judy smiled and sat down on the couch on the other side of the room. "Actually, Dick, the League invited me. Seems they think I have some important things to say."

"The only thing anyone wants to hear you say is 'I'm in way over my head and I quit!'"

"Sorry, Dick, but you can't always get what you want. Oh, but you must know that by now, what with your little problem..."

He shifted uncomfortably in place. "You bring that up tonight and I'll sue your ass off!"

"Well, I could stand to lose a few more pounds off that part of my body, but I don't think there'll be time to talk about your physical, uh, deficiencies. Unless you bring it up first. I've got important points to make."

"Like what? Free beauty make-overs for everyone who visits the capitol?"

"Well, now that you mention it, those eyebrows of yours..."

Dick got up. "I'm outta here," he snarled. "No point getting my blood pressure up over you. You won't change the outcome of this election. The voters are too smart to let someone joke their way into the Governor's mansion."

Judy nodded. "That's exactly what I'm counting on."

Dick stormed out. Judy sat down and studied her cards.

A few seconds later Mark slid in.

"I wanted to make sure you were alone."

"Why, sir? Whatever did you have in mind?"

"What I have in mind will have to wait for tonight at the hotel." He kissed her hard on the lips. "I just came to give you the good news."

"What good news?"

"You've moved up a half percent in the polls."

"Is that good?"

"Honey, that's so good, I bet ol' Dick will pee in his pants when he finds out."

Lynn stuck her head in. "It's time!"

"Okay," Judy said taking a deep breath. To Mark, she whispered, "Remind me when this is all over to show you my appreciation."

****

Once on the soundstage under the hot lights, Judy was thankful for the thinness of her jacket. And the fact that she had chosen *Only in Texas* hairspray. Anything else would have melted.

She sat on a maroon swivel chair next to the democratic candidate. Dick sat in the far chair. All three had note cards in front of them on the table. Judy's were the only ones that were pink.

It only took twelve minutes before one of the panelists, the Channel 2 female anchor, asked Dick the question he'd been dreading, but had most prepared for. "Mr. Sloan, can you lay to rest the growing allegations of improper, if not illegal sexual conduct during high school, college, and at your former business?"

As camera 3 focused on Dick, he put on his best 'Let Me Explain This So You'll Understand' grin. "Well, Dee, if I may call you that, you have to understand that back when I was in high school and college, things were different." Roger cringed. The dick was ad-libbing. "Women always said no and we guys knew that sometimes they meant yes…"

Loud grumbles buzzed through the audience. Jimmy Joe, who hadn't been able to keep himself away, was sitting in the third row, smack in the middle. Listening to Dick's response, he slumped down in his seat like he was trying to disappear. Judy looked past the lights and cameras and noticed looks of hostility on four of the panelists' faces. The fifth, a portly guy with a bad toupee who worked for the *Houston Business News*, looked like he agreed with Dick.

Roger seethed some more. How the hell was he supposed to handle Dick Sloan? Dick was a time bomb that wasn't waiting to go off, he was detonating himself right here on this stage.

Dick must have noticed the reaction his comments had gotten, because he caught himself. "I didn't mean that in a bad way. It's just like... Well, it's kind of like spanking. You know what I mean?" Dick knew he was digging a hole, but he kept on shoveling. "My dad hit me with a belt, and I'm sure a few of yours did too. We didn't call it child abuse back then, we called it parenting."

****

The next day, the political pundits couldn't agree who had won the debate. It was either Buddy Lamar or Judy Parker. They could agree on one thing, however. Dick Sloan was in shit so deep even a pair of deluxe duck hunting hip-waders wouldn't keep it off.

Wrap it Up

It looked like business as usual at Lone Star Cuts. Nan wore an oversized Hawaiian print muumuu and teased a client's hair. Ima applied bluebonnet decals to a 17-year old's newly-applied acrylic nails. Leon flirted, ostensibly with his client at the shampoo station, but really with his own reflection in the mirror. Bernadette swept her station. Donna added up last month's receipts. Judy rolled Maddy Lou's hair on pink perm rollers. Only there was something different about Judy. Perhaps it was the big honking diamond on her finger.

Maddy Lou, she of the eternal tan, was dressed in black spandex, with a dance skirt tied at the waist. She flicked a stray hair from her chin and as she did caught the glimmer from Judy's ring.

"Well, I see you finally got some good jewelry."

Judy flashed it proudly in front of Maddy Lou. "Yep. Mark decided to make an honest woman out of me. Even though he doesn't get to be First Man of Texas."

Maddy Lou expressed the appropriate amount of reverence for the diamond, despite the fact that she was wearing one six times its size. "You'd have made a great governor! Maybe next time."

"Hey, I couldn't have done it without you. All those signatures you collected. But I gotta admit, as much fun as it was, I'm glad it's all over." Judy looked around the salon. "And now things are pretty much back to normal. Except that I'm getting married. Oh, and Sarah is buying me my own cable special."

"You'll have to let me know when that's coming on."

"Sure thing. First I have to write some new jokes." Judy laughed. "Seems like everyone in the state has heard all the old stuff!"

The phone rang. Donna stopped adding and answered it on the second ring. "Lone Star Cuts. Wait just a sec." She looked over towards Judy. "Judy, can you take this?"

Judy took the three steps toward the phone as she excused herself to Maddy Lou. "A rabid fan, obviously."

Donna nodded. "It's the Director of Women's Health Research."

Judy took the phone. "Hey, Gail! How are you?" Gail was the only one affiliated with Judy's campaign who'd actually been bitten by the political bug. And she'd impressed Buddy Lamar so much during the campaign, that he'd appointed her to head his new department at the Texas Department of Health. "You're kidding? That's great! The power of breasts, hon! I'll tell everyone. See you soon. Unless you're too important to hang out with hairdressers and comics now. Okay, next week. See you then."

Judy hung up. "Gail said to tell everyone that Governor Lamar just announced that he's tripling the budget for breast cancer research!"

Maddy Lou stared down at her own ample, and healthy, breasts. "That's great!"

Donna added, "Great!"

Judy smiled. Gail was so happy at her new job. And Lynn was happy too. She was on a two-week bike tour of France. Something Annie had talked her into. The last time Judy had talked to her, she sounded relaxed and very happy. And she promised Judy she was coming back to Austin. With Annie. They were moving into Lynn's place. It was official. Lynn was finally part of a couple. Her mom was so proud.

Suddenly Donna remembered the letter that had come while Judy was out fetching a cup of coffee. "Judy I forgot, this came in the mail today."

Judy squinted at the letter and asked, "What is it?"

"I don't know, but it's from the Austin Independent School District."

Donna got up and went over to Judy and put the envelope in her hand. "Go ahead. Open it." Judy did.

"Well?" Donna asked impatiently.

"I passed! You are now looking at a high school graduate!"

Maddy Lou stood up and hugged Judy. Then she sat back down. "So, Miss High School Graduate, you planning to finish my hair any time soon?"

Judy smiled. "Maybe. If you're good to me." She picked up another pink roller and fingered it gently. "So, Maddy Lou, did I tell you about that time…"

Just then the sounds of a motorcycle roared from the parking lot. Judy looked outside to see Mark sitting on a Harley.

"Who's that?" Maddy Lou asked, noticing Judy's distraction.

"That's my fiancé," enunciating the word slowly. "Turns out he has a little bad boy in him. But just the right amount."

Judy waved at Mark through the window. He blew her a kiss and rode off.

Judy went on with her story. Everyone laughed at the punch lines. Maddy Lou's hair slowly got done. It was just another day at Lone Star Cuts, Home of Texas' Biggest Hair.

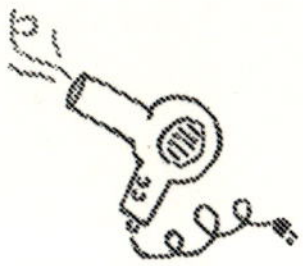

The End

Printed in the United States
202119BV00001B/1-51/A

9 780967 448671